IN MEMORIAM

R.H.F. and R.H.K.

Iain T. Adamson

A General Topology Workbook

Birkhäuser
Boston • Basel • Berlin

Iain T. Adamson
Department of Mathematics and Computer Science
Dundee DD1 4HN
Scotland

Library of Congress Cataloging-in-Publication Data

Adamson, Iain T.
  A general topology workbook / Iain T. Adamson
          p.   cm.
      ISBN 0-8176-3844-X (S : acid free). – ISBN 3-7643-3844-X (S:
  acid-free
        1. Topology.   I.   Title.
  QA611.A346   1995
  514'.322–dc20   95-33319
                                                                  CIP

Printed on acid-free paper
©1996 Birkhäuser Boston

Birkhäuser ®

ISBN 0-8176-3844-X   SPIN 10782858
ISBN 3-7643-3844-X

Typeset by the author in LATEX.
Printed and bound by Quinn-Woodbine, Woodbine, NJ.
Printed in the United States of America.

9 8 7 6 5 4 3 2

Contents

INTRODUCTION

This book has been called a *Workbook* to make it clear from the start that it is not a conventional textbook. Conventional textbooks proceed by giving in each section or chapter first the definitions of the terms to be used, the concepts they are to work with, then some theorems involving these terms (complete with proofs) and finally some examples and exercises to test the readers' understanding of the definitions and the theorems. Readers of this book will indeed find all the conventional constituents—definitions, theorems, proofs, examples and exercises— but not in the conventional arrangement.

In the first part of the book will be found a quick review of the basic definitions of general topology interspersed with a large number of exercises, some of which are also described as theorems. (The use of the word *Theorem* is not intended as an indication of difficulty but of importance and usefulness.) The exercises are deliberately not "graded"—after all the problems we meet in mathematical "real life" do not come in order of difficulty; some of them are very simple illustrative examples; others are in the nature of tutorial problems for a conventional course, while others are quite difficult results. No solutions of the exercises, no proofs of the theorems are included in the first part of the book—this is a *Workbook* and readers are invited to try their hand at solving the problems and proving the theorems for themselves. I have been persuaded, with some reluctance, to offer suggestions about how to tackle the exercises which are not entirely straightforward; really dedicated *Workbook*-ers should ignore these! The second part of the book contains complete solutions to all but the most utterly trivial exercises and complete proofs of the theorems.

It has been widely recognised that general topology is a branch of mathematics particularly well adapted to independent study based on material carefully prepared by a teacher who otherwise gives minimal assistance. The most celebrated practitioner of this method was R.L. Moore (1882-1974) whose success with it is legendary. One factor in Moore's success must have been his insistence on hand-picking his students, preferring those who came to him *tabula rasa*. Few of his imitators and followers have enjoyed the luxury of being able to do this and have had to adapt his method to their more mundane circumstances. This book has grown from my attempts to provide a self-learning introduction to general topology for several generations of students in The University of Dundee, not all of whom would have been selected by Moore—though all of them responded enthusiastically to the method.

My Dundee students were presented, by instalments, with the introductory material and the exercises (though with much less generous hints than appear in the book) and were given the solutions only after the class had met to discuss the problems. Not all the students solved every problem for themselves, but there was a much higher participation rate than in conventional lecture courses. Two students who tackled the course by virtually unsupervised reading had particularly marked success.

In a class situation it is easy to ensure that students do not see the solutions to the exercises before they have tried to solve them; it is not so easy when exercises and solutions appear between the same two covers. But my readers are reminded that this is a *Workbook* and they are warmly invited to work at the exercises before turning to the solutions.

I must put on record my thanks to the Dundee students who have worked so cheerfully with the contents of this book, especially Malcolm Dobson and Ross Anderson. I am also deeply indebted to my friend Keith Edwards for his meticulous assistance, both typographical and mathematical. My gratitude to my wife for her encouragement during the writing and typesetting of this book is beyond measure.

Dundee, Scotland   Iain T. Adamson
August 1995

Part I

EXERCISES

Chapter 1

TOPOLOGICAL SPACES

Throughout the book we use the following standard notation:

N is the set of natural numbers $\{0, 1, 2, \ldots\}$;
Q is the set of rational numbers;
R is the set of real numbers.
For each positive integer n, $\mathbf{R}^n$ is the set of all ordered n-tuples $(x_1, x_2, \ldots, x_n)$ of real numbers.
For each set E the set of all subsets of E is denoted by $\mathbf{P}(E)$; if X is any subset of E its complement with respect to E is denoted by $C_E(X)$.

We begin with a definition which applies to arbitrary sets the notion of distance along a line or in a plane or three-dimensional space familiar from elementary Euclidean geometry. Let E be a set. A **metric** on E is a mapping d from $E \times E$ to **R** such that

D1. For all points x, y in E we have $d(x,y) \geq 0$ and $d(x,y) = 0$ if and only if $x = y$;
D2. For all points x, y in E we have $d(x,y) = d(y,x)$;
D3. For all triples of points x, y, z of E we have

$$d(x,z) \leq d(x,y) + d(y,z).$$

(This is the **Triangle Inequality**).

The first seven exercises are straightforward verifications that the given mappings d satisfy the defining conditions for a metric. The only likely difficulties are in Exercise 2, where to prove the Triangle Inequality we need to use Cauchy's Inequality (if $a_1, \ldots, a_n; b_1, \ldots, b_n$ are real numbers then $(\sum a_i^2)(\sum b_i^2) \geq (\sum a_i b_i)^2)$ and in Exercise 7, where the verification that $d(f, g) = 0$ implies $f = g$ requires us to remember that if a continuous function is nonzero at a point a, then it is nonzero in a subinterval containing a.

Exercise 1. Let $E = \mathbf{R}$, $d(x, y) = |y - x|$ for all x, y in E. Show that d is a metric on E; we call this the **usual metric**.

Exercise 2. Let $E = \mathbf{R}^n, d(x, y) = \sqrt{(\sum_{i=1}^n (y_i - x_i)^2)}$ for all $x = (x_1, \ldots, x_n)$, $y = (y_1, \ldots, y_n)$ in E. Show that d is a metric on E; we call this the **usual metric** or the **Euclidean metric** on E.

Exercise 3. Let $E = \mathbf{R}^2, d(x, y) = |y_1 - x_1| + |y_2 - x_2|$ for all $x = (x_1, x_2), y = (y_1, y_2)$ in E. Show that d is a metric on E.

Exercise 4. Let $E = \mathbf{R}^2, d(x, y) = \max \{|y_1 - x_1|, |y_2 - x_2|\}$ for all $x = (x_1, x_2), y = (y_1, y_2)$ in E. Show that d is a metric on E.

Exercise 5. Let A be a set, E the set of bounded real-valued functions of A, $d(f, g) = \sup_{x \in A} |g(x) - f(x)|$ for all f, g in E. Show that d is a metric on E; we call this the **uniform metric** on E.

Exercise 6. Let E be any set,

$$d(x, y) = \begin{cases} 0 & \text{if } x = y \\ 1 & \text{if } x \neq y. \end{cases}$$

Show that d is a metric on E. This metric is called the **discrete metric** on E.

Exercise 7. Let I be an interval of the real line, E the set of continuous real-valued functions of I, $d(f, g) = \int_I |g - f|$ for all f, g in E. Show that d is a metric on E.

A **metric space** is an ordered pair (E, d) consisting of a set E and a metric d on E. Once the metric d has been fixed for a given discussion, we may talk of "the metric space E" instead of (E, d).

Let (E, d) be a metric space, r a positive real number and a a point of E. Then the set $V_d(a, r) = \{x \in E : d(a, x) < r\}$ is called the r-**ball**

with centre a; we may omit the subscript d if the metric is clear from the context. The set $V'_d(a, r) = \{x \in E : 0 < d(a, x) < r\}$ is called the **deleted** or **punctured** r-**ball with centre** a. (This "ball" terminology is clearly suggested by the Euclidean metric in $\mathbf{R}^3$.)

It is not easy to give a brief motivation for the next definition, in the sense that the first two chapters, if not the whole book, form the real motivation. But the definition was probably suggested by consideration of the example of metric spaces in Exercise 15 with its introduction of "open sets" which made it possible to talk about continuity, for example, without explicitly mentioning the metric.

Let E be a set. A **topology** on E is a subset T of $\mathbf{P}(E)$, i.e. a set of subsets of E, such that

T1. E and $\emptyset$ belong to T;

T2. The union of every family of sets in T is also a set in T;

T3. The intersection of every finite family of sets in T is also a set in T.

Let T and T' be topologies on a set E. We say that T is **finer than** T' and that T' is **coarser than** T if $T \supseteq T'$.

Exercises 8 and 9 are completely trivial verifications of the defining conditions; but the topologies introduced there are frequently referred to. Exercises 10 to 13 are not difficult, but need some care in recognising the different cases that have to be dealt with; the topologies introduced in these exercises are useful as examples and counterexamples, of which every young topologist needs to have a large stock.

Exercise 15, as we mentioned above, is the prototype of topologies on a set; the verification of condition **T3** depends on the fact that the minimum of a finite set of positive real numbers is a positive real number.

Exercise 16 is straightforward. Exercise 17 requires us to find for each point q in $V_d(a, r)$ a ball with centre q entirely included in $V_d(a, r)$. Drawing a little picture of the special situation where $E = \mathbf{R}^2$ and d is the Euclidean metric will suggest the appropriate radius.

In Exercises 18 and 19 we recall that a topology is a set of sets; hence, to prove two topologies T_1 and T_2 equal, we have to proceed by showing first that $T_1 \subseteq T_2$ and then that $T_2 \subseteq T_1$.

Exercise 8. Let E be any set. Show that $T = \mathbf{P}(E)$, the set of all subsets of E, is a topology on E. This topology is called the **discrete topology**. It is clearly the finest topology on E.

Exercise 9. Let E be any set. Show that $T = \{\emptyset, E\}$ is a topology on E. This topology is called the **trivial topology** or the **indiscrete topology**. It is clearly the coarsest topology on E.

Exercise 10. Let E be a set, p a point of E. Show that the collection $T_p = \{\emptyset\} \cup \{X \in \mathbf{P}(E) : p \in X\}$ is a topology on E. Such a topology is called a **particular point topology**. A special case is the **Sierpinski topology** T on $E = \{0, 1\}$ given by $T = \{\emptyset, \{0\}, E\}$.

Exercise 11. Let E be a set, p a point of E. Show that the collection $T_{-p} = \{E\} \cup \{X \in \mathbf{P}(E) : p \notin X\}$ is a topology on E. Such a topology is called an **excluded point topology**.

Exercise 12. Let E be an infinite set. Show that the collection $T = \{\emptyset\} \cup \{X \in \mathbf{P}(E) : C_E(X) \text{ is finite}\}$ is a topology on E; we call this the **finite complement topology** on E.

Exercise 13. Let E be an infinite set, p a point of E. Show that $T = \{X \in \mathbf{P}(E) : p \notin X \text{ or } C_E(X) \text{ is finite}\}$ is a topology on E; this is called the **Fort topology**.

Exercise 14. Find all possible topologies on a set of three elements.

Exercise 15. Let (E, d) be a metric space. A subset U of E is said to be **open relative to the metric** d if, for every point x of U, there is a positive real number r_x such that $V_d(x, r_x) \subseteq U$. Let T_d be the set of all subsets of E which are open relative to d. Show that T_d is a topology on E; we call this the **topology induced by the metric** d.

Exercise 16. Let E be any set. Show that the topology induced on E by the discrete metric is the discrete topology.

Exercise 17. Let (E, d) be a metric space, a a point of E and r a positive real number. Prove that $V_d(a, r) \in T_d$.

Exercise 18. Let d be a metric on a set E; let d' be the mapping from $E \times E$ to $\mathbf{R}$ given by $d'(x,y) = \min\{1, d(x,y)\}$ for all x, y in E. Prove that d' is a metric on E and that $T_{d'} = T_d$.

Exercise 19. Show that the metrics on $\mathbf{R}^2$ defined in Exercises 3 and 4 induce the same topology as the Euclidean metric.

A topology T on a set E is said to be **metrizable** if there exists a metric d on E such that $T = T_d$. (Notice that Exercises 18 and 19 show that the metric d is unlikely to be unique.)

Exercises 20 and 21 introduce relaxations in the definition of *metric* which have proved useful. Exercise 15 provides the model for defining the induced topologies. The verification that the mappings described in Exercises 22 and 23 satisfy the relevant conditions is entirely routine.

Exercise 20. Let E be a set. A **pseudometric** on E is a mapping p from $E \times E$ to $\mathbf{R}$ satisfying the condition

D1p. For all points x, y in E we have $p(x,y) \geq 0$ and if $x = y$ then $p(x,y) = 0$

together with the conditions **D2** and **D3** for a metric. A **pseudometric space** is an ordered pair (E, p) consisting of a set E and a pseudometric p on E.

Show how to define the **topology induced by a pseudometric**.

Exercise 21. Let E be a set. A **quasimetric** on E is a mapping q from $E \times E$ to $\mathbf{R}$ satisfying conditions **D1** and **D3** for a metric but not necessarily condition **D2**. A **quasimetric space** is an ordered pair (E, q) consisting of a set E and a quasimetric q on E.

Show how to define the **topology induced by a quasimetric**.

Exercise 22. Let E be any set, f a mapping from E to $\mathbf{R}$. Show that the mapping p from $E \times E$ to $\mathbf{R}$ given by setting $p(x,y) = |f(x) - f(y)|$ for all x, y in E is a pseudometric on E.

Exercise 23. Show that the mapping q from $\mathbf{R} \times \mathbf{R}$ to $\mathbf{R}$ given by

$$q(x,y) = \begin{cases} y - x & \text{if } x \leq y \\ 1 & \text{otherwise} \end{cases}$$

is a quasimetric on **R**.

Exercise 24. Let (E, p) be a pseudometric space. Consider the relation $R = \{(x, y) \in E \times E : p(x, y) = 0\}$. Show that R is an equivalence relation on E. Show how to define a metric p^* on E/R such that, for all x, y in E, we have $p^*(\eta(x), \eta(y)) = p(x, y)$, where η is the canonical surjection from E onto E/R.

The verification that R is an equivalence relation is straightforward. To define a metric p^* on E/R we must define $p^*(X, Y)$ for all R-classes X, Y; it is natural to do this by choosing elements x, y of E from X, Y respectively and setting $p^*(X, Y) = p(x, y)$. But it is necessary to show that $p^*(X, Y)$ so defined depends only on the R-classes X and Y themselves and not on the choices of x and y.

Exercise 25. Let E be a set. An **ultrametric** on E is a mapping u from $E \times E$ to **R** satisfying conditions **D1** and **D2** for a metric together with

D3u. For all points x, y, z in E we have

$$u(x, z) \leq \max\{u(x, y), u(y, z)\}.$$

Of course an ultrametric is a metric.

Show that every triangle in an ultrametric space is isosceles and that the equal sides are not shorter than the base.

Exercise 26. Let **Q** be the set of rational numbers; let p be a prime number. We define a mapping u from **Q** $\times$ **Q** to **R** as follows. If $x = y$ we set $u(x, y) = 0$; if x and y are distinct rational numbers we write $y - x = p^{\alpha}m/n$ where m and n are relatively prime integers not divisible by p and then set $u(x, y) = p^{-\alpha}$. Show that u is an ultrametric on E; we call it the **p-adic metric**.

A **topological space** is an ordered pair (E, T) consisting of a set E and a topology T on E. We call E the **underlying set** and T the **topology** of the topological space (E, T). When the topology T is clear from the context, we may talk of "the topological space E" instead of (E, T).

Let (E, T) be a topological space. The sets in the collection T are called T-**open sets** or simply open sets if the topology T is clear from the context. When we use this terminology the defining conditions **T1-T3** become:

T1. E and $\emptyset$ are open sets;

T2. The union of every family of open sets is open;

T3. The intersection of every finite family of open sets is open.

We recall that, if T and T' are topologies on a set E, then T is said to be finer than T' and T' coarser than T if $T \supseteq T'$; thus T is finer than T' if and only if every T'-open subset of E is T-open.

Let T be a topology on a set E. A **base** for the topology T is a subset B of T such that every set in T is the union of some family of sets in B.

The point of introducing the notion of a base for a topology is simply to make it possible to describe a topology without actually giving *all* the sets which comprise it: it is sufficient to give the sets of a base.

Exercises 27 and 28 are called *Theorems* not because they are difficult but to signal that the results are important enough to remember. Exercise 27 gives a criterion for a collection of subsets of a set E to be a base for a given topology T on E; the proof is direct. Exercise 28 gives a condition for a collection of subsets to be a base for *some* topology, so that here we have to show how to construct the topology from the alleged base. Clearly, if B is a base for a topology T, then all the sets of T are unions of families of sets in B; so it is not unreasonable to examine the set of all such unions and see whether this is a topology.

Exercises 29 to 31 are routine verifications that the conditions of Exercise 28 are satisfied; again they provide additions to our stock of examples of topologies.

Theorem 1 = Exercise 27. Let T be a topology on a set E, B a subset of T. Then B is a base for T if and only if for every set U in T and every point x of U there exists a set W in B such that $x \in W \subseteq U$.

Theorem 2 = Exercise 28. Let E be a set, B a set of subsets of E such that (1) $\bigcup B = E$ and (2) for every pair of sets W_1, W_2 in B and every point x of $W_1 \cap W_2$ there exists a set W in B such that $x \in W \subseteq W_1 \cap W_2$. Then there exists a unique topology T_B on E of which B is a base.

Exercise 29. Let E be a set, $(X_i)_{i \in I}$ a family of pairwise disjoint subsets of E such that $\bigcup_{i \in I} X_i = E$. Prove that $X = \{X_i\}$ is a base for a topology T_X on E. Such a topology is known as a **partition topology**. A particular example is the **odd-even topology** on $\mathbf{N}$ (the set of natural numbers) for which the base is $\{\{2k, 2k+1\}\}_{k \in \mathbf{N}}$.

Exercise 30. Let E be a set which is totally ordered by a relation denoted by $\leq$ (this means that for all elements x, y of E we have either $x \leq y$ or $y \leq x$). Show that E together with the collection of all open intervals, i.e. all sets of the form $\{t \in E : a < t < b\}$ together with those of the form $\{t \in E : t < a\}$ and those of the form $\{t \in E : t > a\}$, is a base for a topology on E; this is called the **order topology** on E.

Exercise 31. Let E be a set which is totally ordered by a relation denoted by $\leq$. Show that E, together with the collection of all sets of the form $\{t \in E : t > a\}$, is a base for a topology on E; we call this the **right order topology** on E.

Let E be a set, S a set of subsets of E. The set $\tau(S)$ of all topologies on E which include S is clearly non-empty since $\mathbf{P}(E)$ belongs to $\tau(S)$. Let $T(S)$ be the intersection of $\tau(S)$, i.e. the set of all sets each of which belongs to all the topologies in $\tau(S)$. Then it is easily verified that $T(S)$ is a topology on E which includes S; it is clearly the smallest topology on E which does so. We call $T(S)$ the **topology induced on E by S** or the **topology on E generated by S**. The advantage of the ideas introduced here is that we can be even more economical in our description of a topology than by giving all its sets or even all the sets in a base. Exercise 32 describes all the sets of the topology generated by a given set of subsets.

Theorem 3 = Exercise 32. Let E be a set, S a set of subsets of E. Then the topology $T(S)$ on E generated by S consists of the set E together with all subsets which are unions of families of intersections of finite families of sets in S.

Let (E, T) be a topological space. A subset F of E is said to be T-**closed** (or simply **closed**) if $C_E(F) \in T$. Closed sets have the following properties:

C1. E and $\emptyset$ are closed sets;

C2. The intersection of every family of closed sets is a closed set;

C3. The union of every finite family of closed sets is a closed set.

Exercise 33 shows that, instead of describing a topology directly by giving all its sets (the open sets of the topology), we may instead prescribe which are to be the closed sets of the topology (they will, of course, have to satisfy the conditions for sets closed relative to a topology). Obviously the (open) sets of the topology must be the complements of the proposed closed sets.

Exercise 34 is a simple exercise on complements of unions and intersections. (A **countable set** is one which is either finite or in one-to-one correspondence with the set of natural numbers; a family $(U_i)_{i \in I}$ is countable if its index set I is countable.)

Exercises 35 and 36 need care but no great inspirational insights.

Theorem 4 = Exercise 33. Let E be a set, K a collection of subsets of E such that

(1) E and $\emptyset$ belong to K;

(2) the intersection of every family of sets in K belongs to K;

(3) the union of every finite family of sets in K belongs to K.

Then there is a unique topology T on E such that K is the collection of T-closed sets of E.

Exercise 34. Let (E, T) be a topological space. A subset of E is called a G_δ-set if it is the intersection of a countable family of T-open sets; a subset of E is called an F_σ-set if it is the union of a countable family of T-closed subsets. Prove:

(1) the complement in E of a G_δ-set is an F_σ-set and *vice versa*;

(2) if $K = \bigcup_{n \in \mathbb{N}} K_n$ (where all the sets K_n are T-closed) is an F_σ-set there is a family $(F_n)_{n \in \mathbb{N}}$ of closed subsets of E such that $K = \bigcup_{n \in \mathbb{N}} F_n$ and $F_n \subseteq F_{n+1}$ for all natural numbers n.

Exercise 35. Let (E, T) be a topological space; let p be a point which does not belong to the set E. Let $E^* = E \cup \{p\}$ and define the collection $T^* = \{U \cup \{p\} : U \in T\} \cup \{\emptyset\}$. Show that T^* is a topology on E^* and that the T^*-closed subsets of E are precisely the T-closed subsets of E. We call T^* the **closed extension topology** of T. Show that the particular point topology T_p of Exercise 10 is the closed extension of the discrete topology on $C_E\{p\}$.

Exercise 36. Let $E = \{x \in \mathbf{R} : -1 \leq x \leq 1\}$, $T = \{X \in \mathbf{P}(E) :$ either $0 \notin X$ or $(-1,1) \subseteq X\}$. Show that T is a topology on E, called the **either-or topology**. Prove that the only T-closed subsets of E are $\{1\}, \{-1\}, \{1, -1\}$ and all subsets which contain 0.

Let (E, T) be a topological space, a a point of E. A subset V of E is called a T-**neighbourhood** (or simply a **neighbourhood**) of a if there exists a T-open set U such that $a \in U$ and $U \subseteq V$. If V is a T-neighbourhood of a then we write $V' = C_V\{a\}$; we call V' the **punctured** or **deleted** T-**neighbourhood** corresponding to V. By a **fundamental system of** T-**neighbourhoods** of a or a T-**neighbourhood base** of a we mean a collection $BN(a)$ of neighbourhoods of a such that, for every T-neighbourhood V of a, there exists a set W in $BN(a)$ such that $W \subseteq V$.

No special comments are needed for Exercises 37 to 39; the result of Exercise 38 is constantly used—its usefulness, rather than the difficulty of its proof, justifies calling it a Theorem.

Exercise 37. Let (E, d) be a metric space, p a point of E. Show that the set of balls with centre p and rational radii forms a neighbourhood base for p (for the topology induced by the metric d).

Theorem 5 = Exercise 38. Let (E, T) be a topological space. A subset U of E is T-open if and only if it is a T-neighbourhood of each of its points.

Exercise 39. Let T and T' be topologies on a set E. Prove that T is finer than T' if and only if, for every point x of E, every T'-neighbourhood of x is a T-neighbourhood of x.

Let (E, T) be a topological space, p a point of E and $N(p)$ the set of T-neighbourhoods of p. Then it follows from the definition of neighbourhoods that $N(p)$ satisfies the following:

N1. Every subset of E which includes a set in $N(p)$ is itself in $N(p)$;

N2. The intersection of each finite family of subsets in $N(p)$ belongs to $N(p)$;

N3. The point p belongs to each member of $N(p)$;

N4. For every subset V of E in $N(p)$ there exists a subset W in

$N(p)$ such that V is a T-neighbourhood of every point in W.

The verification of conditions **N1** to **N3** is immediate; **N4** follows from the observation that if $V \in N(p)$, i.e. V is a T-neighbourhood of p, then V includes a T-open set W containing p—hence in $N(p)$—which, by Theorem 5, is a T-neighbourhood of each of its points; hence V, which includes W, is a T-neighbourhood of all the points of W by condition **N1**.

Exercise 40 shows that a topology may be defined on a set E by prescribing for each point its neighbourhoods with respect to the topology (provided, of course, the proposed neighbourhood collections in fact satisfy the conditions for neighbourhood collections with respect to a topology). Referring to Theorem 5, we see that the (open) sets in the topology must be those which belong to the proposed neighbourhood collections for each of their points. It is not hard to verify that the collection of all such sets satisfies the defining conditions for a topology and that each neighbourhood of a point x relative to this topology belongs to the given set $N(x)$. It is not so easy to show conversely that every set V in the given proposed neighbourhood set $N(x)$ is actually a neighbourhood of x in the topology defined in this way. Condition (4) is required for this purpose, and perhaps also a judicious glance at the answer!

Theorem 6 = Exercise 40. Let E be a set; let $(N(x))_{x \in E}$ be a family of non-empty sets of subsets of E such that

(1) for each point x of E every subset of E which includes a subset in $N(x)$ belongs to $N(x)$;

(2) for each point x of E the intersection of each finite family of sets in $N(x)$ belongs to $N(x)$;

(3) for each point x of E the point x is in every subset in $N(x)$;

(4) for each point x of E and each set V in $N(x)$ there exists a set W in $N(x)$ such that V belongs to $N(y)$ for every point y in W.

Show that there exists a unique topology T on E such that for each point x of E the set $N(x)$ is the set of T-neighbourhoods of x.

Let (E, T) be a topological space, A a subset of E.

The **interior** of A is the union of the set of all T-open subsets of E which are included in A, i.e. the largest T-open subset included in A. We denote it by $\operatorname{Int}_T(A)$ or simply $\operatorname{Int} A$.

The **closure** of A is the intersection of the set of all T-closed subsets

of E which include A, i.e. the smallest T-closed subset which includes
A. We denote it by $\mathrm{Cl}_T(A)$ or simply $\mathrm{Cl}\,A$.

Exercises 41 and 42 follow quite easily from the definitions of interior
and closure. (We recall that, in order to prove that two sets E_1 and
E_2 are equal, we may proceed by showing first that $E_1 \subseteq E_2$ and then
that $E_2 \subseteq E_1$.) In Exercise 43 (where $\subset$ denotes proper inclusion) the
family we are looking for must clearly be infinite (it follows easily by
mathematical induction from Exercise 41 (1) that for a finite family
$(A_i)_{i \in I}$ we have $\mathrm{Int}(\bigcap_{i \in I} A_i) = \bigcap_{i \in I} \mathrm{Int}(A_i)$).

Exercise 44 asks us to show that a topology may be defined on a set
E by prescribing for each subset its closure in the topology (provided,
of course, the proposed association of closures to subsets satisfies the
properties of a closure operation in a topology). It is clear from the
definition of closure that if A is T-closed then $\mathrm{Cl}_T(A) = A$. So the
closed sets for the alleged topology in which κ is the closure operation
must be the subsets X such that $\kappa(X) = X$: these comments should
explain the definition of the collection T_κ. In proving that this collection
is a topology, it is useful to notice that it follows from condition (3)
that if X and Y are subsets of E such that $X \subseteq Y$ then we have
$\kappa(X) \subseteq \kappa(Y)$.

Exercises 45 to 48 are all relatively uncomplicated.

Exercise 41. Let (E, T) be a topological space, A and B subsets
of E. Prove
   (1) $\mathrm{Int}(A \cap B) = \mathrm{Int}\,A \cap \mathrm{Int}\,B$;
   (2) $\mathrm{Cl}(A \cup B) = \mathrm{Cl}\,A \cup \mathrm{Cl}\,B$;
   (3) If $A \subseteq B$ then $\mathrm{Cl}\,A \subseteq \mathrm{Cl}\,B$ and $\mathrm{Int}\,A \subseteq \mathrm{Int}\,B$.

Exercise 42. Let (E, T) be a topological space, A a subset of E.
Prove that $\mathrm{Cl}\,(C_E(A)) = C_E(\mathrm{Int}\,A)$.

Exercise 43. Give an example of a topological space (E, T) and a
family of subsets $(A_i)_{i \in I}$ of E for which $\mathrm{Int}(\bigcap_{i \in I} A_i) \subset \bigcap_{i \in I} \mathrm{Int}(A_i)$.

Exercise 44. Let E be a set, κ a mapping from $\mathbf{P}(E)$ to $\mathbf{P}(E)$
such that
   (1) for every subset X of E we have $\kappa(X) \supseteq X$;
   (2) for every subset X of E we have $\kappa(\kappa(X)) = \kappa(X)$;
   (3) for all subsets X and Y of E we have $\kappa(X \cup Y) = \kappa(X) \cup \kappa(Y)$;
   (4) $\kappa(\emptyset) = \emptyset$.

Let $T_\kappa = \{X \in \mathbf{P}(E) : \kappa(C_E(X)) = C_E(X)\}$. Show that T_κ is a topology on E and that we have $\mathrm{Cl}\,_{T_\kappa}(X) = \kappa(X)$ for every subset X of E.

Exercise 45. Let E be an infinite set; let κ be the mapping from $\mathbf{P}(E)$ to $\mathbf{P}(E)$ given by

$$\kappa(X) = \begin{cases} X & \text{if } X \text{ is a finite subset of } E \\ E & \text{if } X \text{ is an infinite subset of } E. \end{cases}$$

Show that κ satisfies the conditions of Exercise 44 and that the topology T_κ is the finite complement topology of Exercise 12.

Let (E, T) be a topological space, A a subset of E. The **frontier** of A is defined to be the closed set $\mathrm{Fr}\, A = \mathrm{Cl}\, A \cap \mathrm{Cl}\,(C_E(A))$.

A is said to be **everywhere dense** (or simply **dense**) if $\mathrm{Cl}\, A = E$.

A is said to be **nowhere dense** if $\mathrm{Int}(\mathrm{Cl}\, A) = \emptyset$.

A is said to be **meagre** if it is the union of a countable family of nowhere dense sets; meagre sets are also referred to as **sets of the first category**. Sets which are not of the first category are said to be **sets of the second category**. (E, T) is said to be a **first (second) category space** if E is a first (second) category subset of E according to the definitions above; first category spaces are also called **meagre spaces**.

Exercise 46. Let E be a non-empty set, T the discrete topology on E, A any subset of E. Find $\mathrm{Int}\, A$, $\mathrm{Cl}\, A$, $\mathrm{Fr}\, A$. Prove that (E, T) is a second category space.

Exercise 47. Let E be a non-empty set, T the trivial topology on E, A any subset of E. Find $\mathrm{Int}\, A$, $\mathrm{Cl}\, A$, $\mathrm{Fr}\, A$. Prove that (E, T) is a second category space.

Exercise 48. Let T be the right order topology on $\mathbf{R}$. Show that for every real number r the set $P_r = \{x \in \mathbf{R} : x < r\}$ is nowhere dense in $\mathbf{R}$. Deduce that $(\mathbf{R}, T)$ is meagre.

Let (E, T) be a topological space, A a subset of E, x a point of E.

Then x is said to be an **interior point** of A if A is a T-neighbourhood of x.

We say that x is an **adherent point** or **point of closure** of A if every T-neighbourhood of x meets A (i.e. has non-empty intersection with A).

x is a **cluster point** or **accumulation point** of A if every deleted T-neighbourhood of x meets A.

x is an **ω-accumulation point** of A if every T-neighbourhood of x contains infinitely many points of A.

x is a **condensation point** of A if every T-neighbourhood of x contains uncountably many points of A.

x is a **frontier point** of A if every T-neighbourhood of x meets both A and $C_E(A)$.

A point a of A is called an **isolated point** of A if it is not a cluster point of A.

Exercises 49 and 50 give us useful alternative descriptions of the interior and closure of a set; their proofs are direct, each of them (of course) in two parts. The ease with which we can handle Exercises 51 and 52 depends on which of the alternative descriptions we choose. Exercises 53 to 58 follow quite quickly from the definitions of the terms involved, though Exercise 55 asks us to exercise some ingenuity in constructing examples, so that we have to choose appropriate spaces to work in.

Theorem 7 = Exercise 49. The points of Int A are precisely the interior points of A.

Theorem 8 = Exercise 50. The points of Cl A are precisely the adherent points of A.

Exercise 51. Let (E, T) be a topological space, X any subset of E. Prove that Cl $X = X \cup$ Fr X and Int $X = C_X(\text{Fr } X)$

Exercise 52. Let (E, T) be a topological space, X any subset of E. For every subset X of E set $\alpha(X) = \text{Int (Cl } X)$. Prove:

(1) if $X \subseteq Y$ then $\alpha(X) \subseteq \alpha(Y)$;

(2) if X is open then $X \subseteq \alpha(X)$;

(3) for every subset X of E we have $\alpha(\alpha(X)) = \alpha(X)$;

(4) if X and Y are disjoint open sets then $\alpha(X)$ and $\alpha(Y)$ are also disjoint.

Exercise 53. Let (E, T) be a topological space. A subset X of E is said to be **regular open** if $X = \alpha(X)$ (as in Exercise 45). Prove that the intersection of each finite family of regular open subsets of E is regular open. Show that the sets $X = \{x \in \mathbf{R} : 0 < x < \frac{1}{2}\}$ and

$Y = \{x \in \mathbf{R} : \frac{1}{2} < x < 1\}$ are regular open subsets of $\mathbf{R}$ with its usual topology, but that $X \cup Y$ is not regular open.

Exercise 54. Let (E, T) be a topological space, A and B subsets of E such that $E = A \cup B$. Prove that $E = \operatorname{Cl} A \cup \operatorname{Int} B$.

Exercise 55. Let (E, T) be a topological space, A and B subsets of E. Prove that
  (1) $\operatorname{Fr}(\operatorname{Cl} A) \subseteq \operatorname{Fr} A$ and $\operatorname{Fr}(\operatorname{Int} A) \subseteq \operatorname{Fr} A$;
  (2) $\operatorname{Fr}(A \cup B) \subseteq \operatorname{Fr} A \cup \operatorname{Fr} B$.
Give an example where the three sets in (1) are distinct. Give an example where the inclusion in (2) is proper.

Exercise 56. Let (E, T) be a topological space, D a dense subset and U an open subset of E. Prove that $U \subseteq \operatorname{Cl}(D \cap U)$.

Exercise 57. Let (E, T) be a topological space, A a subset of E. Show that A meets every dense subset D of E if and only if $\operatorname{Int} A$ is non-empty.

Exercise 58. Let E be any set, T_p the particular point topology on E determined by the point p. Prove that $\operatorname{Cl}\{p\} = E$. Show that, if F is any proper closed subset of E, then $\operatorname{Int} F = \emptyset$. Show that, if X is any subset of E which contains p and t is any point distinct from p, then t is a cluster point but not an ω -accumulation point of X.

Let (E, T) be a topological space. The topology T and the space (E, T) are said to be **first countable** if every point of E has a countable T-neighbourhood base. First countable topologies and spaces are also said to satisfy the **first axiom of countability**. The topology T and the space (E, T) are said to be **second countable** if there exists a countable base for T. Second countable topologies and spaces are also said to satisfy the **second axiom of countability**.

The space (E, T) is said to be **separable** if E has a countable dense subset.

For those who proceed systematically, the following exercises are all relatively mechanical consequences of the definitions. The only one which may cause difficulty is Exercise 61, where we have to construct a countable base for a separable metric space. Here we should obviously begin with the countable dense subset D provided by the separability condition. The metric then allows us to look at balls centred at the points of D, which are open sets (by Exercise 17) and to choose countably many of these (say the balls with radii $1/n$ for all positive integers n). The collection of balls so obtained is countable and can be shown to be a base.

Exercise 59. Prove that every second countable space is separable.

Exercise 60. Prove that the following spaces are separable but not second countable:
   (1) E any uncountable set, T any particular point topology on E.
   (2) E any uncountable set, T the finite complement topology on E.

Exercise 61. Let (E, d) be a metric space, T the topology induced by the metric d. Prove that if (E, T) is separable then it is second countable.

Exercise 62. Prove that every second countable space is first countable.

Exercise 63. Prove that the following topological spaces (E, T) are first countable but not second countable:
   (1) E any uncountable set, T the discrete topology on E.
   (2) E any uncountable set, T any particular point topology on E.

Exercise 64. Show that the following topological spaces (E, T) are second countable:
   (1) E any countable set, T the discrete topology on E.
   (2) E any set, T the trivial topology on E.
   (3) $E = \mathbf{N}$, T the odd-even topology.
   (4) E any countable set, T any particular point topology on E.
   (5) E any countable set, T any excluded point topology on E.
   (6) $E = \mathbf{R}$, T the usual metric topology.
   (7) $E = \mathbf{R}$, T the right order topology.

Chapter 2

MAPPINGS OF TOPOLOGICAL SPACES

Let f be a mapping from a set E to a set E'. If A is a subset of E we use the notation $f^{\rightarrow}(A)$ for the direct image of the subset A under f, i.e. the set consisting of the images $f(a)$ of all points a in A. Again, if A' is a subset of E', we use $f^{\leftarrow}(A')$ to denote the inverse image of A' under f, i.e. the set of all points x of E for which the image $f(x)$ lies in A'.

Let (E, T) and (E', T') be topological spaces, x a point of E and f a mapping from E to E'. We say that f is (T, T')-**continuous at** x (or simply **continuous at** x) if for every T'-neighbourhood V' of $f(x)$ there exists a T-neighbourhood V of x such that $f^{\rightarrow}(V) \subseteq V'$. We say that the mapping f is (T, T')-**continuous** (or simply **continuous**) if it is (T, T')-continuous at every point x of E.

Theorem 1 = Exercise 65. Let (E, T) and (E', T') be topological spaces, f a mapping from E to E'. Then the following conditions are equivalent:

 (1) f is continuous;

 (2) for every T'-open subset U' of E' the inverse image $f^{\leftarrow}(U')$ is T-open;

 (3) for every T'-closed subset F' of E' the inverse image $f^{\leftarrow}(F')$ is T-closed.

The second condition in Exercise 65 is often taken as the definition of continuity, emphasizing that in general topology (in contrast to real analysis) we are more concerned with "global" properties of functions than "local" properties.

Theorem 2 = Exercise 66. Let $(E_1, T_1), (E_2, T_2), (E_3, T_3)$ be topological spaces; let f and g be mappings from E_1 to E_2 and E_2 to E_3 respectively. If f is (T_1, T_2)-continuous and g is (T_2, T_3)-continuous then $g \circ f$ is (T_1, T_3)-continuous.

Exercise 66 depends on the (easily established) set-theoretic result that $(g \circ f)^{\leftarrow}(U_3) = f^{\leftarrow}(g^{\leftarrow}(U_3))$ for each subset U_3 of E_3.

Let (E, T) and (E', T') be topological spaces. A mapping f from E to E' is called a (T, T')-**homeomorphism** or a **topological transformation** if
  (1) f is bijective (so that it has an inverse mapping from E' to E which we denote by f^{-1}),
  (2) f is (T, T')-continuous,
  (3) f^{-1} is (T', T)-continuous.
Two topological spaces (E, T) and (E', T') are said to be **homeomorphic** if there exists a (T, T')-homeomorphism from E to E'. A property $P(X)$ is called a **topological property** if whenever $P((E, T))$ holds and (E', T') is homeomorphic to (E, T) then $P((E', T'))$ also holds.

Theorem 3 = Exercise 67. Let (E, T) and (E', T') be topological spaces. A mapping f from E to E' is a (T, T')-homeomorphism if and only if it is (T, T')-continuous and there is a (T', T)-continuous mapping g from E' to E such that $f \circ g = I_{E'}$ and $g \circ f = I_E$ (where $I_E, I_{E'}$ are the identity mappings from E to E, E' to E' respectively).

Again let (E, T) and (E', T') be topological spaces, f a mapping from E to E'. Then f is said to be (T, T')-**open** if the direct image under f of each T-open subset of E is a T'-open subset of E'. Similarly f is said to be (T, T')-**closed** if the direct image under f of each T-closed subset of E is a T'-closed subset of E'.

Theorem 4 = Exercise 68. Let (E, T), (E', T') be topological spaces, f a bijection from E onto E'. Then f is a (T, T')-homeomorphism if and only if it is (T, T')-continuous and either (T, T')-open or (T, T')-closed.

Exercise 68 depends essentially on the observation (easily verified) that if f is a bijection from E to E', with inverse f^{-1}, then for each subset X of E we have $(f^{-1})^{\leftarrow}(X) = f^{\rightarrow}(X)$.

Exercise 69. Let $E = \mathbf{R}^2, E' = \mathbf{R}$ and consider the mapping from E to E' given by $f(x, y) = x$ for all points (x, y) of E. Prove that f is (T, T')-continuous but not (T, T')-closed (where T and T' are the usual metric topologies).

The second part of Exercise 69 requires us to produce an example of a closed set in the coordinate plane whose projection on the x-axis is not closed.

Exercise 70. Let (E, T) and (E', T') be topological spaces, f a mapping from E to E'. Prove that f is (T, T')-continuous if and only if, for every subset X' of E', we have $f^{\leftarrow}(\mathrm{Cl}_{T'}(X')) \supseteq \mathrm{Cl}_T(f^{\leftarrow}(X'))$.

Exercise 70 consists of two parts. In the first we suppose that f is (T, T')-continuous and try to show that for every subset X' of E' we have $f^{\leftarrow}(\mathrm{Cl}_{T'}(X')) \supseteq \mathrm{Cl}_T(f^{\leftarrow}(X'))$. To do this it is helpful to recall that for any two subsets X', Y' of E' such that $X' \subseteq Y'$ we have $f^{\leftarrow}(X') \subseteq f^{\leftarrow}(Y')$, and that the T-closure of a subset of E is the smallest T-closed set which includes it. For the second part of the exercise, we must show that if $f^{\leftarrow}(\mathrm{Cl}_{T'}(X')) \supseteq \mathrm{Cl}_T(f^{\leftarrow}(X'))$ for all subsets X' of E' then f is (T, T')-continuous. To do this we should clearly be thinking of condition (3) of Exercise 65; so we start with a T'-closed subset of E' and try to prove that its inverse image under f is T-closed. The trick here is to remember that the closure of a closed subset is the subset itself.

Exercise 71. Let (E, T) and (E', T') be topological spaces, f a mapping from E to E'. Prove that f is (T, T')-open if and only if for every subset X of E we have $f^{\rightarrow}(\mathrm{Int}_T(X)) \subseteq \mathrm{Int}_{T'}(f^{\rightarrow}(X))$.

Exercise 71 is tackled in a similar way to Exercise 70: we use the fact that if X and Y are subsets of E such that $X \subseteq Y$, then $f^{\rightarrow}(X) \subseteq f^{\rightarrow}(Y)$; we recall also that the interior of a subset is the largest open set included in it and that the interior of an open subset is the subset itself.

Exercise 72. Let (E, T) and (E', T') be topological spaces, f a bijection from E onto E'. Show that f is a (T, T')-homeomorphism if and only if T' is the finest topology T_0 on E' such that f is (T, T_0)-continuous.

In the first part of Exercise 72 we suppose that f is a (T, T')-homeomorphism. Then, if we have a topology T_0 on E' such that f is (T, T_0)-continuous, we have to show that $T_0 \subseteq T'$. To do this we take a T_0-open set U_0 and use the (T, T_0)-continuity of f and the (T', T)-continuity of f^{-1} to show that $U_0 \in T'$. In the second part, in order to show that f^{-1} is (T', T)-continuous, we proceed by contradiction: we suppose that there is a T-open set U whose inverse image under f^{-1} is not T'-open and use this to construct a topology T_0 finer than T' such that f is (T, T_0)-continuous.

Exercise 73. Let (E, T) be a topological space. Prove that (1) T is the discrete topology on E if and only if for every topological space (E', T') every mapping f from E to E' is (T, T')-continuous; (2) T is the trivial topology if and only if for every topological space (E', T') every mapping f from E' to E is (T', T)-continuous.

In Exercise 73 the "only if" parts are immediate. To establish the "if" parts we apply the stated condition to a particular choice of function f: the appropriate identity function will do the job.

Exercise 74. Let (E, T) be a topological space, f and g (T, T')-continuous mappings from E to $\mathbf{R}$ (with its usual topology). Prove that $\{x \in E : f(x) = g(x)\}$ is a closed subset of E. Deduce that if $f(x) = g(x)$ for all points x of a dense subset D of E then $f(x) = g(x)$ for all points x of E.

Chapter 3

INDUCED AND COINDUCED TOPOLOGIES

In the first part of this chapter we show how a family of mappings from a set E to the underlying sets of a family of topological spaces may be used to construct a topology on E. There are two important special cases, described in Examples 1 and 2, where we apply the general construction to define subspace and product topologies.

Let E be a set, $((E_i, T_i))_{i \in I}$ a family of topological spaces, and $(f_i)_{i \in I}$ a family of mappings from E to the family $(E_i)_{i \in I}$. (By this we mean, of course, that for each index i in I f_i is a mapping from E to E_i.) Let $\mathbf{A}$ be the set of all topologies T on E such that, for every index i in I, the mapping f_i is (T, T_i)-continuous. This set is non-empty since the discrete topology belongs to it. The intersection of $\mathbf{A}$ (i.e. the collection of sets U such that U is in every one of the topologies in $\mathbf{A}$) is again a topology and is easily seen to belong to $\mathbf{A}$; it is clearly the coarsest topology T on E such that each mapping f_i is (T, T_i)-continuous. We call it the **topology induced** on E by the family of mappings $(f_i)_{i \in I}$.

Theorem 1 = Exercise 75. In the situation described, the topology induced on E by $(f_i)_{i \in I}$ is the topology $T(S)$ generated by the set $S = \{X \in \mathbf{P}(E) : \text{for some index } i \text{ in } I \text{ there is a set } U_i \text{ in } T_i \text{ such that } X = f_i^{\leftarrow}(U_i)\}$.

Exercise 75 gives us a way of getting our hands on the sets in the induced topology. The proof is as usual in two parts. We show that

$T(S)$ is included in the induced topology T_0 by noting that each of the sets in S belongs to T_0 since all the mappings f_i are (T_0, T_i)-continuous. Then we show that T_0 is included in $T(S)$ by noticing that $T(S)$ belongs to the family **A**.

Theorem 2 = Exercise 76. Still in the same situation, let (E', T') be a topological space and g a mapping from E' to E. Then g is (T', T)-continuous if and only if each mapping $f_i \circ g$ is (T', T_i)-continuous.

In Exercise 76 if g is (T', T)-continuous then each $f_i \circ g$ is (T', T_i)-continuous by Exercise 66. If all the mappings $f_i \circ g$ are (T', T_i)-continuous, then we prove that g is (T', T)-continuous by showing that for each set U in T we have $g^{\leftarrow}(U)$ in T'. To do this we use the result of Exercise 75 and the description in Exercise 32 of the sets in $T = T(S)$.

Example 1. Let (E, T) be a topological space, A a subset of E; let i be the inclusion mapping of A in E given by $i(a) = a$ for all points a of A. The topology induced on A by the family consisting of the single mapping i is called the **subspace topology** or **relative topology** on A; we denote it by T_A. The topological space (A, T_A) is called a **subspace** of (E, T).

In this case the set S described in Theorem 1 is $S = \{X \in \mathbf{P}(A) :$ there exists an open set U in T such that $X = i^{\leftarrow}(U) = A \cap U\}$. Clearly in this case S is itself already a topology on A. Hence T_A consists of all subsets of A of the form $A \cap U$ with U in the topology T. The sets in the topology T_A are sometimes said to be **open in** A.

Exercises 77 to 81 are all reasonably simple consequences of the definitions, though Exercise 79 may require a little ingenuity.

Exercise 77. Let (E, T) be a topological space, A a subset of E, T_A the subspace topology on A. Prove (1) a subset B of A is T_A-closed if and only if $B = A \cap F$ where F is a T-closed subset of E; (2) for every subset B of A we have $\mathrm{Cl}_{T_A}(B) = A \cap \mathrm{Cl}_T(B)$; (3) for every subset B of A we have $\mathrm{Int}_T(B) \subseteq \mathrm{Int}_{T_A}(B)$. Give an example where the inclusion in (3) is proper.

Exercise 78. Let (E, T) be a topological space, D a dense subset of E, a any point of D and V a T_D-neighbourhood of a. Prove that $\mathrm{Cl}_T(V)$ is a T-neighbourhood of a.

Exercise 79. Let (E,T) be a topological space, A and B subsets of E such that $E = A \cup B$. Let M be a subset of $A \cap B$ which is both T_A-open and T_B-open. Prove that M is T-open.

Exercise 80. Let (E,T) be a separable topological space. Prove that if U is a T-open subset of E then (U, T_U) is also separable.

Exercise 81. Let E be an uncountable set, p a point of E and T_p the particular point topology on E determined by p. Let $A = C_E\{p\}$. Prove that (E, T_p) is separable but that $(A, (T_p)_A)$ is not.

Example 2. Let $((E_i, T_i))_{i \in I}$ be a family of topological spaces. Let $E = \prod_{i \in I} E_i$ and, for each index i in I, let π_i be the projection mapping from E to E_i. The topology induced on E by the family of mappings $(\pi_i)_{i \in I}$ is called the **product topology** on E; we denote it by $\prod_{i \in I} T_i$. The topological space $(\prod_{i \in I} E_i, \prod_{i \in I} T_i)$ is called the **topological product** of the family $((E_i, T_i))_{i \in I}$.

The product topology is generated by the collection of sets of the form $\pi_i^{\leftarrow}(U_i)$ where $i \in I$ and $U_i \in T_i$, i.e. sets of the form $\prod_{i \in I} X_i$ where $X_i = E_i$ for all indices i but one, say i_0, and $X_{i_0} \in T_{i_0}$. Then a base for the product topology is provided by the collection of all intersections of finite families of such sets, i.e. by the collection of sets of the form $\prod_{i \in I} X_i$ where $X_i \in T_i$ for all indices i in I and $X_i = E_i$ for all but a finite set of indices i in I.

Exercise 82. Let $((E_i, T_i))_{i \in I}$ be a family of non-empty topological spaces. If U is any $(\prod T_i)$-open subset of $\prod E_i$, prove that $\pi_i^{\rightarrow}(U) = E_i$ for all but finitely many indices i in I.

Exercise 82 follows easily from the fact that U is a union of basic open sets as just decribed.

Exercise 83. Let $((E_i, T_i))_{i \in I}$ be a family of topological spaces, (E, T) its topological product. For each index i in I let A_i be a subset of E_i. Prove that $\mathrm{Cl}_T(\prod A_i) = \prod(\mathrm{Cl}_{T_i}(A_i))$.

The proof here proceeds in two parts. First we show that if x is any point of $\mathrm{Cl}_T(\prod A_i)$ then, for each index j in I, the projection $\pi_j(x)$ is a T_j-adherent point of A_j. Next we prove that, for each point x of $\prod \mathrm{Cl}_{T_j} A_j$ and each basic T-open set U containing x, each projection $\pi_j^{\rightarrow}(U)$ of U meets A_j and so U meets $\prod A_j$.

Exercise 84. Let $((E_i, T_i))_{i \in I}$ be a family of topological spaces. Prove that if I is countable and each of the spaces (E_i, T_i) is either (1) separable or (2) second countable then so is the product $(\prod E_i, \prod T_i)$.

The separability part of Exercise 84 is a little delicate. We obviously start with a countable dense subset D_i of each E_i and try to construct a countable dense subset D of $\prod E_i$. Since I is countable there is no loss of generality in thinking of I as the set of natural numbers. Then the elements of $\prod E_i$ can be thought of as "vectors" $(x_0, x_1, \ldots, x_n, \ldots)$ where $x_i \in E_i$ for all i in I. We define D to be the set of all vectors $t = (d_0, d_1, \ldots)$ where each d_i is in the corresponding countable dense set D_i and there is a natural number m such that, for all indices i from a certain point $i(t)$ onwards, each d_i is the m-th element of the corresponding set D_i. Then with some ingenuity we can prove that D is a countable dense subset of E.

Let $((E_i, T_i))_{i \in I}$ be a family of topological spaces, (E, T) a topological space; for each index i in I, let f_i be a (T, T_i)-continuous mapping from E to E_i. Then we may define a mapping F_I from E to the product $\prod_{i \in I} E_i$ by setting $f_I(x) = (f_i(x))_{i \in I}$ for all points x of E. It follows from Theorem 2 that f_I is $(T, \prod T_i)$-continuous.

Exercise 85. In the situation just described prove

(1) f_I is injective if and only if the family $(f_i)_{i \in I}$ distinguishes points, i.e. if for every pair of distinct points x, y of E there is an index i in I such that $f_i(x) \neq f_i(y)$;

(2) if T_I is the subspace topology induced by $\prod T_i$ on the image of f_I in $\prod E_i$, then f_I is (T, T_I)-open if the family $(f_i)_{i \in I}$ distinguishes points and closed sets, i.e. if for every T-closed subset F of E and every point x of E not in F there is an index i in I such that $f_i(x) \notin \mathrm{Cl}_{T_i}(f_i^{\rightarrow}(F))$.

Assertion (1) is straightforward. To prove assertion (2) let U be a T-open subset of E; to show that $f_I^{\rightarrow}(U)$ is T_I-open we use Theorem 5 of Chapter 1. So let $t = f_I(x)$ be any point of $f_I^{\rightarrow}(U)$. Apply the "distinguishes points and closed sets" condition to $F = C_E(U)$ and x.

Exercise 86. Let $X = \{a, b, c\}$ and define T_0 to be the topology $\{\emptyset, \{a\}, X\}$ on X. Prove that every topological space (E, T) is homeomorphic to a subspace of the topological product (P, T_P) of the family $((X_i, T_i))_{i \in I}$ where $I = E \cup T$ and for each i in I we have $(X_i, T_i) = (X, T_0)$.

Apply Exercise 85 using the family $(f_i)_{i \in I}$ of mappings from E to $X_i = X$ given by

$$f_p(t) = \begin{cases} c & \text{if } t \neq p \\ b & \text{if } t = p \end{cases}$$

for each p in E and

$$f_U(t) = \begin{cases} a & \text{if } t \in U \\ b & \text{if } t \notin U \end{cases}$$

for each U in T. (You must show that each of these mappings is (T, T_0)-continuous.)

In the second part of the chapter we show how a family of mappings to a set E from the underlying sets of a family of topological spaces may be used to construct a topology on E. The most important example is the case of the quotient topology on a set of equivalence classes.

We let E be a set, $((E_i, T_i))_{i \in I}$ a family of topological spaces; for each index i in I, let g_i be a mapping from E_i to E. Let **A** be the set of all topologies T on E such that each of the mappings g_i is (T_i, T)-continuous. Clearly **A** is non-empty since the trivial topology on E belongs to it. The finest topology in **A** is called the **topology coinduced** on E by the family $(g_i)_{i \in I}$.

Exercises 87 and 88 are straightforward.

Theorem 3 = Exercise 87. In the situation described, the topology T coinduced on E by the family $(g_i)_{i \in I}$ consists of all subsets U of E such that $g_i^{\leftarrow}(U) \in T_i$ for every index i in I.

Theorem 4 = Exercise 88. Still in the situation described, let (E', T') be a topological space, f a mapping from E to E', T the topology coinduced on E by $(g_i)_{i \in I}$. Then f is (T, T')-continuous if and only if each mapping $f \circ g_i$ is (T_i, T')-continuous.

Example. Let (E, T) be a topological space, R an equivalence relation on E, η the canonical surjection from E onto E/R (given by $\eta(x) = R$-class of x for each element x of E). The topology coinduced on E/R by the family consisting of the single mapping η is called the **quotient topology** on E/R; it is denoted by T/R. We call $(E/R, T/R)$ the **quotient space** of (E, T) with respect to R.

Let $(E, T), (E', T')$ be topological spaces, f a (T, T')-continuous mapping from E to E'. Let R_f be the equivalence relation on E defined by setting $(x, y) \in R_f$ if and only if $f(x) = f(y)$; let η be the canonical surjection from E onto E/R_f. Let f^* be the canonical bijection from E/R_f onto $B = f^{\rightarrow}(E)$ and j the canonical injection from B to E', so that $f = j \circ f^* \circ \eta$. Then η is $(T, T/R_f)$-continuous and j is $((T')_B, T')$-continuous. Furthermore, since $f^* \circ \eta$ is $(T, (T')_B)$-continuous, it follows from Theorem 4 that f^* is $(T/R_f, (T')_B)$-continuous.

If R is an equivalence relation on a set E and X is a subset of E the R-**saturate** of X is the union of all the R-classes of members of X; so if η is the canonical surjection of E onto E/R the R-saturate of X is the set $\eta^{\leftarrow}(\eta^{\rightarrow}(X))$. A subset X of E is said to be R-**saturated** if it coincides with its saturate. We notice that if $\bar{X}$ is any subset of E/R then $\eta^{\leftarrow}(\bar{X})$ is R-saturated.

Exercise 89 follows from the definition of the quotient topology T/R_f and the comments we have just made about saturated sets. Both directions of Exercise 90 are straightforward. Exercise 91 is best handled by proving the three implications $(1) \implies (2)$, $(2) \implies (3)$, $(3) \implies (1)$.

Theorem 5 = Exercise 89. With the notation just described, f^* is a $(T/R_f, (T')_B)$-homeomorphism if and only if, for every R_f-saturated T-open subset U of E, we have $f^{\rightarrow}(U) \in T_B$.

Let (E, T) be a topological space, R an equivalence relation on E. The relation R is said to be **open** or **closed** if the canonical surjection η from E to E/R is $(T, T/R)$-open or $(T, T/R)$-closed respectively.

Theorem 6 = Exercise 90. Let (E, T) be a topological space, R an equivalence relation on E. Then R is open if and only if, for every T-open set U, the R-saturate of U is also open.

Exercise 91. Let (E, T) be a topological space, R an equivalence relation on E. Show that the following conditions are equivalent:

(1) R is an open relation;

(2) the interior of each R-saturated subset of E is also R-saturated;

(3) the closure of each R-saturated subset of E is also R-saturated.

Theorem 7 = Exercise 92. Let (E, T) be a topological space, R an equivalence relation on E. Then R is closed if and only if for every R-class C and every T-open set U which includes C there exists an R-saturated T-open set W such that $C \subseteq W \subseteq U$.

Theorem 7 consists of two parts. To prove the first part we follow the path

U is T-open $\implies C_E(U)$ is T-closed $\implies \bar{F} = \eta^{\to}(C_E(U))$ is T/R-closed $\implies C_{E/R}(\bar{F})$ is T/R-open $\implies W = \eta^{\leftarrow}(C_{E/R}(\bar{F}))$ is T-open and R-saturated.

For the second part, suppose the condition of the theorem is satisfied and let F be a T-closed subset of E. We must try to show that $G = \eta^{\leftarrow}(\eta^{\to}(F))$ is T-closed. This will follow if we can show that $C_E(G)$ is a T-neighbourhood of each of its points x. To do this we apply the condition of the theorem to the case where $C = R(x)$ and $U = C_E(F)$.

In Exercise 93 we may proceed by looking for an R-class C and an open set U including it with no saturated set between C and U. Exercise 94 is uncomplicated.

Exercise 93. Let $E = \{(x, y) \in \mathbf{R}^2 : y \geq 0\}$. Define an equivalence relation R on E by setting $((x, y), (x', y')) \in R$ if and only if either (1) $y = y' \neq 0$ or (2) $y = y' = 0$ and $x = x'$. Prove that R is not closed. (The topology on E is the subspace topology induced by the Euclidean topology on $\mathbf{R}^2$.)

Exercise 94. Let (E, T) be a topological space, R a closed equivalence relation on E. Let U be a T-open subset of E. According to Theorem 7 we know that, for every R-class C included in U, there is an R-saturated T-open set U_C such that $C \subseteq U_C \subseteq U$. Prove that $\bigcup_{C \subseteq U} C = \bigcup_{C \subseteq U} U_C$.

Exercise 95. Deduce from Exercise 94 that if R is a closed equivalence relation on E and X is an R-saturated subset of E, then the subspace topology induced by (T/R) on the subset $\eta^{\rightarrow}(X)$ coincides with the quotient topology of T_X coinduced by the restriction of the relation R to the subset $X \times X$, that is to say the relation

$$R' = \{(x_1, x_2) \in X \times X : (x_1, x_2) \in R\}.$$

For this exercise we show first that if $\bar{A}$ is a set of the subspace topology induced by T/R on $\eta^{\rightarrow}(X)$, then $\eta^{\leftarrow}(\bar{A})$ belongs to the subspace topology induced by T on X; from this it follows readily that $\bar{A} \in T_X/R'$.

To prove the reverse inclusion, we take a set $\bar{A}$ in T_X/R'; then $\eta^{\leftarrow}(\bar{A})$ is in T_X and hence has the form $U \cap X$ where U is T-open. We let $\mathbf{K}$ be the set of all R-classes included in U and let $B = \bigcup \mathbf{K}$; applying Exercise 94 we see that B is T-open and $\eta^{\rightarrow}(B)$ is T/R-open. We then prove that $\eta^{\leftarrow}(\bar{A}) = X \cap B$ and deduce that $\bar{A} = \eta^{\rightarrow}(X) \cap \eta^{\rightarrow}(B)$ and so is in the subspace topology on $\eta^{\rightarrow}(X)$.

Chapter 4

CONVERGENCE

Let E be a set. A **filter** on E is a non-empty collection F of subsets of E such that

F1. Every subset of E which includes a set of F is itself in F;

F2. The intersection of each finite family of sets in F is in F;

F3. All the sets in F are non-empty.

Let F be a filter on a set E. A collection B of subsets of E is called a **base** for the filter F if (1) $B \subseteq F$ and (2) for every set V in F, there is a set W in B such that $W \subseteq V$; we say also that B **generates** F.

Exercises 96 to 101 either involve routine checking that the conditions of the definitions are satisfied or else are immediate consequences of the definitions.

Exercise 96. Let E be a set, A any non-empty subset of E. Show that the set of subsets of E which include A is a filter on E and that $\{A\}$ is a base for this filter.

Exercise 97. Let (E, T) be a topological space, a any point of E. Prove that the set of all T-neighbourhoods of a is a filter on E (the **T-neighbourhood filter at a**) and that each T-neighbourhood base at a is a base for this filter.

Exercise 98. Let E be an infinite set. Show that the collection of subsets of E which have finite complement is a filter on E. If $E = \mathbf{N}$ this filter is called the **Fréchet filter**. Show that the set of subsets of $\mathbf{N}$ of the form $\{n \in \mathbf{N} : n \geq k\}$ (for all natural numbers k) is a base for the Fréchet filter.

Exercise 99. Let F and G be filters on a set E. Show that a subset X of E belongs to both F and G if and only if there are sets P in F and Q in G such that $X = P \cup Q$.

Theorem 1 = Exercise 100. Let B be a collection of subsets of a non-empty set E. Then B is a base for a filter on E if and only if (1) the intersection of each finite family of sets in B includes a set in B and (2) B is non-empty and $\emptyset$ does not belong to B.

Let A be a collection of subsets of a set E; let A' be the collection of intersections of all finite families of sets in A. If A' does not contain the empty set $\emptyset$, then it satisfies the conditions of Theorem 1 and hence is a base for a filter F on E. We call F the **filter generated by A**.

Exercise 101. Let F and G be filters on a set E. Suppose that for every pair of subsets X, Y of E in $F \cup G$ we have $X \cap Y \neq \emptyset$. Show that the filter generated by $F \cup G$ consists of all sets of the form $P \cap Q$ where $P \in F$ and $Q \in G$.

Let E be a set, R a relation on E (i.e. a subset of $E \times E$); R is called an **order relation** (sometimes a **partial order**) if it is reflexive, antisymmetric and transitive, that is, if

(1) for all $x \in E$ we have $(x, x) \in R$;

(2) for all x, $y \in E$ such that $(x, y) \in R$ and $(y, x) \in R$ we have $x = y$;

(3) for all x, y, $z \in E$ such that $(x, y) \in R$ and $(y, z) \in R$ we have $(x, z) \in R$.

A subset X of E is said to be **totally ordered** by R if, for every pair of elements x, y of X, we have either $x = y$ or $(x, y) \in R$ or $(y, x) \in R$.

E is said to be **inductively ordered by** R if every subset X of E which is totally ordered by R has an R-supremum, i.e. if there is an element m of E such that $(x, m) \in R$ for all x in X and $(m, m') \in R$ for every element m' of E such that $(x, m') \in R$ for all x in X.

Zorn's Lemma states that every inductively ordered set E has an R-maximal element, i.e. an element a of E such that there is no element

x of E for which $(a, x) \in R$ except a itself. Zorn's Lemma is equivalent to the **Axiom of Choice** which states that the product of every family of non-empty sets is non-empty. In this book we accept the validity of the Axiom of Choice and so may make use of Zorn's Lemma.

Theorem 2 = Exercise 102. The set of all filters on a non-empty set E is inductively ordered by inclusion.

To prove Theorem 2 we must show that each set **F** of filters which is totally ordered by inclusion has a supremum. The union of **F** can be shown to satisfy the conditions of Theorem 1; the filter which it generates is then the required supremum.

It follows from Theorem 2 by the application of Zorn's Lemma that the collection of filters on a non-empty set E has ($\subseteq$)-maximal elements: these maximal filters are called **ultrafilters**. It is also easy to show that for every filter F on a set E there is an ultrafilter on E which includes F.

Theorem 3 = Exercise 103. Let F be an ultrafilter on a set E. If A and B are subsets of E such that $A \cup B \in F$ then either $A \in F$ or $B \in F$.

To prove Theorem 3, suppose we have $A \cup B \in F$ but $A \notin F$ and $B \notin F$. Let F' be the collection of all subsets X of E such that $A \cup X \in F$ and show that F' is a filter which properly includes the ultrafilter F, which is impossible.

Theorem 4 = Exercise 104. Let **A** be a collection of subsets of a non-empty set E which generates a filter F on E. If for every subset X of E we have either $X \in \mathbf{A}$ or $C_E(X) \in \mathbf{A}$, then **A** is an ultrafilter on E.

To prove Theorem 4, suppose F' is an ultrafilter which includes F and show that if X is any set in F' then $C_E(X) \notin \mathbf{A}$ and hence $X \in \mathbf{A}$.

Corollary. Let E be a non-empty set, a any element of E. Then the filter consisting of all subsets of E which contain a is an ultrafilter on E.

Exercise 105. Let F be an ultrafilter on a set E. Show that $\bigcap F$ contains at most one point and that if $\bigcap F = \{a\}$, then F is the ultrafilter consisting of all the subsets of E which contain a.

(Suppose $\bigcap F$ contains two distinct points a and b; let F' be the filter generated by $F \cup \{a\}$ and show that F' properly includes the filter F, which is a contradiction.)

Exercise 106. Let A be a subset of a set E; let F be a filter on E. Let F_A be the set of subsets of A of the form $A \cap X$ where X is in F. Show that F_A is a filter on A if and only if all these sets are non-empty. If F is an ultrafilter on E show that this condition is satisfied if and only if A belongs to F and that, when this happens, F_A is an ultrafilter on A.

Theorem 5 = Exercise 107. Every filter F on a non-empty set E is the intersection of the family of ultrafilters which include F.

For each set X not in F show that X fails to belong to an ultrafilter which includes the filter generated by $F \cup \{C_E(X)\}$.

Theorem 6 = Exercise 108. Let f be a mapping from a set E to a set E'. If B is a base for a filter on E then $f^{\rightarrow}(B) = \{f^{\rightarrow}(X)\}_{X \in B}$ is a base for a filter on E'. If B is a base for an ultrafilter on E then $f^{\rightarrow}(B)$ is a base for an ultrafilter on E'.

The proof of the first assertion is straightforward using the criterion of Theorem 1.

To prove the second assertion, suppose that B is base for an ultrafilter F on E and let F' be the filter on E' generated by $f^{\rightarrow}(B)$. Show that F' is an ultrafilter by using Theorem 4 (let X' be any set in F' and consider the two cases $f^{\leftarrow}(X') \in F$ and $f^{\leftarrow}(X') \notin F$).

Theorem 7 = Exercise 109. Let f be a mapping from a set E to a set E'. If B' is a base for a filter on E' then $f^{\leftarrow}(B') = \{f^{\leftarrow}(X')\}_{X' \in B'}$ is a base for a filter on E if and only if every set in B' meets $f^{\rightarrow}(E)$.

In one direction (the *only if* part) this is almost immediate; the *if* part follows easily by using Theorem 1.

Now let (E, T) be a topological space, F a filter on E. A point x of E is said to be a **limit** or a **limit point** of the filter F and F is said to **converge to** x or to be **convergent to** x if the T-neighbourhood filter $V(x)$ at x is included in the filter F. If B is a filter base on E then x is a limit point of B and B converges to x if the filter generated by B converges to x.

Exercise 110. Let T and T' be topologies on a set E. Show that T is finer than T' if and only if every filter F on E which converges to a point a for the topology T also converges to a for the topology T'.

For the *only if* part, recall that if T is finer than T' then every T'-neighbourhood of a point will also be a T-neighbourhood. In the *if* part, we prove that every T'-open set is T-open by showing that it is a T-neighbourhood of each of its points a; we do this by taking the filter F to be the T-neighbourhood filter of a.

Exercise 111. Let (E, T) and (E', T') be topological spaces, f a mapping from E to E', and p a point of E. Prove that if f is (T, T')-continuous at p then, for every filter F on E which converges to p, the filter base $f^{\rightarrow}(F)$ converges to $f(p)$.

This follows easily from the definition of continuity.

Again let (E, T) be a topological space, F a filter on E. A point x is said to be an **adherent point** of F if x is an adherent point of every set in F. The **adherence** of F, Adh F, is the set of all adherent points of F; so Adh $F = \bigcap_{X \in F} \mathrm{Cl}\, X$. If B is a base for a filter on E, x is said to be an adherent point of B if it is an adherent point of the filter based on B. The **adherence** of B, Adh B, is the set of its adherent points.

Exercise 112. Show that Adh $B = \bigcap_{X \in B} \mathrm{Cl}\, X$.

To do this prove that both sides of the equation are equal to the adherence of the filter generated by B.

Exercise 113. Let (E, T) be a topological space, A a subset of E. Prove that a point x of E is adherent to A if and only if there is a filter F on E such that $A \in F$ and F converges to x.

If x is adherent to A then $V_T(x) \cup \{A\}$ generates a filter with the required property. The converse is immediate when we recall the definition of convergence.

Theorem 8 = Exercise 114. Let (E, T) be a topological space, B a base for a filter on E. Let x be a point of E, N a fundamental system of T-neighbourhoods of x. Then (1) x is a limit point of B if and only if every set in N includes a set in B; (2) x is an adherent point of B if and only if every set in N meets every set in B.

All four parts of this Theorem follow by routine applications of the definitions. The Corollaries are simple consequences.

Corollary 1 = Exercise 115. A point x is adherent to a filter F if and only if there is a filter F' which includes F and converges to x.

Corollary 2 = Exercise 116. Every limit point of a filter F is adherent to F.

Corollary 3 = Exercise 117. Every adherent point of an ultra-filter U is a limit point of U.

Let E be a set, F a filter on E, (E', T') a topological space and f a mapping from E to E'. A point x' of E' is called a **limit point** of f **relative to** F and f is said to **converge** to x' **relative to** F if x' is a limit point of the filter base $f^{\rightarrow}(F)$; x' is called an **adherent point** of f **relative to** F if it is an adherent point of the filter base $f^{\rightarrow}(F)$.

Exercise 118. Let $((E_i, T_i))_{i \in I}$ be a family of topological spaces, (E, T) its topological product. Let A be a set, F a filter on A and f a mapping from A to E. Show that f converges to a point x of E relative to F if and only if each coordinate mapping $\pi_i \circ f$ converges to $\pi_i(x)$ relative to F.

Suppose f converges to x relative to F. To show that $\pi_i \circ f$ converges to $\pi_i(x)$ relative to F, we must show that every T_i-neighbourhood V_i of $\pi_i(x)$ includes a set of the form $(\pi_i \circ f)^{\rightarrow}(X)$ with X in F. We do this by using V_i to construct a T-neighbourhood of x in E.

For the converse we take a T-neighbourhood V of x and try to find a set X in F such that $f^{\rightarrow}(X) \subseteq V$. According to the definition of the product topology T the neighbourhood V includes a product of T_i-open sets U_i only finitely many of which are not the whole space E_i. We use the convergence of $\pi_i \circ f$ for the indices i corresponding to this finite collection to produce sets X_i in F such that $(\pi_i \circ f)^{\rightarrow}(X_i) \subseteq U_i$ and show that their (finite) intersection is the set X we want.

Theorem 9 = Exercise 119. In the notation introduced before Exercise 118, x' is a limit point of f relative to F if and only if, for every T'-neighbourhood V' of x', we have $f^{\leftarrow}(V') \in F$; x' is an adherent point of f relative to F if and only if, for every T'-neighbourhood V' of x' and every set X in F, the intersection $V' \cap f^{\rightarrow}(X)$ is non-empty.

The four assertions in the Theorem follow in a straightforward way from the definitions of the terms involved, making use of Theorem 8.

Example 1. Let $E = \mathbf{N}$, the set of natural numbers, F the Fréchet filter on $\mathbf{N}$, f a mapping from $\mathbf{N}$ to E' (so that f is a **sequence** of points of E'). Then x' is a limit point of this sequence relative to the Fréchet filter if and only if, for every T'-neighbourhood V', of x' we have $f^{\leftarrow}(V') \in F$, i.e. $C_{\mathbf{N}}(f^{\leftarrow}(V'))$ is finite. Hence x' is a limit point of f relative to F if and only if, for every T'-neighbourhood V' of x', there exists a natural number k such that $C_{\mathbf{N}}(f^{\leftarrow}(V')) \subseteq [0, k]$, i.e. such that for all natural numbers n greater than k we have $f(n) \in V'$.

Example 2. Let (E, T), (E', T') be topological spaces, f a mapping from E to E'. Let x be a point of E and let $F = V(x)$ be the T-neighbourhood filter at x. Then a point x' of E' is a limit point of f relative to $V(x)$ if and only if, for every T'-neighbourhood V' of x', we have $f^{\leftarrow}(V') \in V(x)$, i.e. if and only if there exists a T-neighbourhood V of x such that $V \subseteq f^{\leftarrow}(V')$ and so $f^{\rightarrow}(V) \subseteq V'$. The limit points of f relative to $V(x)$ are also called the (T, T')-**limit points** of f at x.

Theorem 10 = Exercise 120. In the notation of Example 2, f is (T, T')-continuous at x if and only if $f(x)$ is a (T, T')-limit point of f at x.

An ordered set $(D, \leq)$ is called a **directed set** if every couple $\{x, y\}$ of elements of D has an upper bound.

Let $(D, \leq)$ be a directed set, E any set. A mapping ν from D to E is called a **net in E with domain D**.

Let ν be a net in E with domain D, A a subset of E. We say that ν is **eventually in** A if there exists an element k in D such that $\nu(n) \in A$ for all elements n in D such that $n \geq k$. We say that ν is **frequently in** A if for every element n of D there exists an element n' of D such that $n' \geq n$ and $\nu(n') \in A$.

Let D and D' be directed sets, ν and ν' nets in E with domains D and D' respectively. We say that ν' is a **subnet** of ν if there exists a mapping φ from D' to D such that

(1) $\nu' = \nu \circ \varphi$ and

(2) for every element n in D there is an element n' in D' such that $\varphi(d') \geq n$ for all elements d' in D' such that $d' \geq n'$.

Example. Let $(D, \leq)$ be the set of natural numbers $\mathbf{N}$ with its usual ordering; $(D, \leq)$ is clearly a directed set. Let E be any set. Then a sequence s of points of E is a net in E with domain $\mathbf{N}$. A second sequence s' of points of E (which is also a net in E with domain $\mathbf{N}$) is a subsequence of s if s' is a subnet of s.

Theorem 11 = Exercise 121. Let ν be a net in E with domain D. Let $\mathbf{A}$ be a collection of subsets of E such that ν is frequently in every subset in $\mathbf{A}$. If the intersection of each pair of members of $\mathbf{A}$ includes a member of $\mathbf{A}$, then there exists a subnet ν' of ν such that, for every member X of $\mathbf{A}$, we have ν' eventually in X.

To construct a subnet of ν, we must first choose a directed set $\bar{D}$ as its domain. One possible approach is to look at $D' = D \times \mathbf{A}$ and show that the relation R' on D' given by setting $((d, A), (d_1, A_1)) \in R'$ if and only if $d \leq d_1$ and $A \supseteq A_1$ is an order relation; if we define $\bar{D} = \{(d, A) \in D' : \nu(d) \in A\}$ and $\bar{R}$ is the restriction of R' to $\bar{D}$, then it is not hard to show that $(\bar{D}, \bar{R})$ is a directed set. Now let $\nu' = \nu \circ \varphi$ where φ is the mapping from $\bar{D}$ to D given by $\varphi((d, A)) = d$ for all (d, A) in $\bar{D}$. It is now fairly easy to prove that ν' is a subnet of ν and that it is eventually in every set of $\mathbf{A}$.

Let (E, T) be a topological space, x a point of E. Let ν be a net in E with domain D. Then ν is said to **converge** to x and x is said to be a **limit point** of ν if, for every T-neighbourhood V of x, the net ν is eventually in V. The point x is said to be an **adherent point** of ν if, for every T-neighbourhood V of x, the net ν is frequently in V.

Theorem 12 = Exercise 122. Let (E, T) be a topological space, ν a net in E. A point a of E is an adherent point of ν if and only if there is a subnet of ν which converges to a.

If a is an adherent point of ν the set of T-neighbourhoods of a satisfies the conditions on **A** in Theorem 11. If a is not an adherent point of ν then there must be a T-neighbourhood V of a such that ν is not frequently in V, so that ν (and hence all its subnets) must be eventually in $C_E(V)$.

Let ν be a net in a set E. Let

$$F(\nu) = \{X \in \mathbf{P}(E) : \nu \text{ is eventually in } X\}.$$

Then $F(\nu)$ is a filter on E, which we call the **filter associated with the net** ν.

Let F be a filter on a set E. Then F is directed by the order relation R defined by setting $(X, Y) \in R$ if and only if $X \supseteq Y$. Let ν be any mapping from F to E such that $\nu(X) \in X$ for every set X in F. Then ν is a net in E; we say that it is **associated with the filter** F.

Theorem 13 = Exercise 123. Let (E, T) be a topological space, F a filter on E and x a point of E. Then F converges to x if and only if every net associated with F converges to x.

We notice that if F converges to x then, for every T-neighbourhood V of x, we have $V \in F$; then, for any set X of F which is R-greater than V, we have $\nu(X) \in V$.

If F does not converge to x there must be a T-neighbourhood V of x which does not belong to F; so every set X in F meets $C_E(V)$. This observation allows us to construct a net associated with F which is not eventually in V and so does not converge to x.

Theorem 14 = Exercise 124. Let (E, T) be a topological space, ν a net in E and x a point of E. Then ν converges to x if and only if the filter $F(\nu)$ converges to x.

This follows at once from the definitions of the terms involved.

Theorem 15 = Exercise 125. Let $(E, T), (E', T')$ be topological spaces, f a mapping from E to E' and x a point of E. Then f is (T, T')-continuous at x if and only if for every net ν in E which converges to

x the net $f \circ \nu$ in E' converges to $f(x)$.

The *only if* part is routine. The *if* part is handled by supposing that f is not continuous at x, so that there is a T'-neighbourhood V' of $f(x)$ such that, for every T-neighbourhood V of x, the set $f^{\rightarrow}(V)$ is not included in V'; this allows us to define a net ν with domain $V_T(x)$ which converges to x but is such that $f \circ \nu$ does not converge to $f(x)$.

Exercise 126. Let ν be a net in a set E with domain D; let $\nu' = \nu \circ \varphi$ be a subnet of ν with domain D'. Prove (1) if ν converges to a point a of E then so does ν'; (2) if a is an adherent point of ν' then a is an adherent point of ν.

These results follow readily from the definitions of the terms involved.

A net ν in a set E is called an **ultranet** or a **universal net** if for every subset X of E we have either ν eventually in X or ν eventually in $C_E(X)$.

Exercise 127. Prove (1) every subnet of an ultranet is an ultranet; (2) every net has a subnet which is an ultranet.

The first part follows by an argument similar to that used in the first part of Exercise 126.

In the second part, given a net $\nu : D \to E$, we have to produce a subnet which is an ultranet. To do this we use Theorem 11: we consider the collection **S** of all sets Q of subsets of E such that (1) ν is frequently in every subset in Q and (2) Q is closed under finite intersection. It is easy to show that **S** is inductively ordered by the inclusion relation and so, by Zorn's Lemma, has a maximal element Q_0. By Theorem 11, ν has a subnet ν' such that for every member X of Q_0 we have ν' eventually in X.

To show that ν' is an ultranet, we must show that for every subset X of E we have either ν' eventually in X or ν' eventually in $C_E(X)$. We do this by showing that for every subset A of E either A or its complement belongs to Q_0; to do this we must use the $(\subseteq)$-maximal property of Q_0.

Exercise 128. Let E be a set, ν a net in E and F a filter on E. Prove (1) if ν is an ultranet then the filter $F(\nu)$ associated with ν is an ultrafilter; (2) if F is an ultrafilter then every net associated with F is an ultranet.

The first part follows from the definitions of ultranet and $F(\nu)$ using Theorem 4.

To prove the second part, let ν be any net associated with an ultrafilter F. We must show that for every subset Y of E we have either (1) ν is eventually in Y or (2) ν is eventually in $C_E(Y)$. Suppose (1) does not hold; then ν is certainly frequently in $C_E(Y)$. We can show that $F \cup \{C_E(Y)\}$ generates a filter including F, which must therefore be F since F is an ultrafilter. So $C_E(Y) \in F$. We deduce that for every set X' in F such that $X' \subseteq C_E(Y)$ we have $\nu(X') \subseteq C_E(Y)$, i.e. ν is eventually in $C_E(Y)$.

Chapter 5

SEPARATION AXIOMS

Let (E, T) be a topological space.

(1) The topology T and the space (E, T) are said to be **T_0** if, for every pair of distinct points x, y of E, there exists a T-neighbourhood of one which does not contain the other, i.e. either a T-neighbourhood of x which does not contain y or a T-neighbourhood of y which does not contain x. A T_0 space is sometimes called a **Kolmogorov space**.

Exercise 129. Show that every particular point topology on a set is T_0.

To tackle this let a and b be distinct points of E and consider separately the cases where (1) one of these points is the particular point p defining the topology and (2) a and b are both distinct from p.

Exercise 130. Let (E, T) be a topological space. Prove that T is T_0 if and only if, for every pair of distinct points x, y of E, we have $\mathrm{Cl}_T\{x\} \neq \mathrm{Cl}_T\{y\}$.

For the *only if* part show that we have either $x \in \mathrm{Cl}_T\{x\}$ but $x \notin \mathrm{Cl}_T\{y\}$ or $y \in \mathrm{Cl}_T\{y\}$ but $y \notin \mathrm{Cl}_T\{x\}$. (Remember the criterion in Exercise 50 for a point to be in the closure of a set.)

For the *if* part it is best to proceed by contradiction: suppose that the condition holds but that (E, T) is not T_0.

Exercise 131. Let (E, T) be a topological space. Let R be the relation on E defined by setting $(x, y) \in R$ if and only if $\mathrm{Cl}_T\{x\} = \mathrm{Cl}_T\{y\}$. Show that the quotient space $(E/R, T/R)$ is T_0.

To prove this result, show first that the canonical surjection η from E onto E/R is an open mapping (to do this let U be any T-open subset of E and prove that $\eta^{\leftarrow}(\eta^{\rightarrow}(U)) = U$). Now let $X = \eta(x)$ and $Y = \eta(y)$ be distinct points of E/R; since X and Y are distinct, it follows that $(x, y) \notin R$ and so $\mathrm{Cl}_T\{x\} \neq \mathrm{Cl}_T\{y\}$; so there is a T-open set U containing one of x and y but not the other. Show that $\eta^{\rightarrow}(U)$ contains the corresponding one of X and Y but not the other.

Exercise 132. Show that the topology T_p induced by a pseudometric p on a set E is T_0 if and only if p is actually a metric.

Let (E, T) be a topological space. Let A be the relation defined on E by setting $(x, y) \in A$ if and only if $x \in \mathrm{Cl}_T\{y\}$.

Exercise 133. Show that the relation A is reflexive and transitive. Show also that A is antisymmetric if and only if the topology T is T_0.

A topological space (E, T) is called an **Alexandrov space** if the intersection of every family of T-open subsets of E is T-open.

Exercise 134. Let (E, T) be an Alexandrov space. Show that a subset K of E is T-closed if and only if whenever $y \in K$ and $(x, y) \in A$ (where A is the relation defined before Exercise 133) we have $x \in K$.

(2) The topology T and the space (E, T) are said to be $\mathbf{T_1}$ if, for every pair of distinct points x, y of E, there exists a T-neighbourhood of each which does not contain the other, i.e. a T-neighbourhood of x which does not contain y and a T-neighbourhood of y which does not contain x. A T_1 space is sometimes called a **Fréchet space**.

Exercise 135. Let E be a set with more than one point. Show that every particular point topology on E is T_0 but not T_1.

Exercise 136. The topology T on $\mathbf{Z}$, the set of integers, generated by the set of subsets of the form $\{2n-1, 2n, 2n+1\}$ for all integers n, is called the **digital topology** on $\mathbf{Z}$. Show that if k is an odd integer then $\{k\}$ is T-open, while if k is an even integer then $\{k\}$ is T-closed. Prove that the digital topology is T_0 but not T_1.

Exercise 137. Show that the topology T_q induced by a quasimetric on a set E is T_1.

Theorem 1 = Exercise 138. Let (E, T) be a topological space. The following conditions are equivalent:
(1) T is a T_1 topology;
(2) For every point x of E the set $\{x\}$ is T-closed;
(3) For every point x of E the intersection of the T-neighbourhood filter of x is $\{x\}$.

Prove the implications in the order (1) $\implies$ (2), (2) $\implies$ (3), (3) $\implies$ (1); all three are straightforward.

Exercise 139. Let (E, T) be a T_1 space, A a subset of E and x an adherent point of A. Prove that if x does not belong to A, then every T-neighbourhood of x contains infinitely many points of A.

Suppose we had a T-neighbourhood V of x containing only finitely many points of A. If we apply the T_1 property to x and these finitely many points, then we can construct a neighbourhood of x which does not meet A.

Exercise 140. Let (E, T) be a T_1 space with a finite base for its topology T. Show that E is finite and T is discrete.

Start with a point x of E which must lie in a set B_1 of the base. If $B_1 \neq \{x\}$ we apply the T_1 property to produce another set B_2 of the base containing x and properly included in B_1. Proceeding in this way, we have a strictly decreasing sequence of sets of the base, which must terminate since the base is finite.

Exercise 141. Prove that the only T_1 topology on a finite set is the discrete topology.

(3) The topology T and the space (E, T) are said to be **T$_2$** if, for every pair of distinct points x and y of E, there exist disjoint T-neighbourhoods of x and y. T_2 spaces are nearly always called **Hausdorff spaces**, except in France, where they are called **espaces séparés**.

Exercise 142. Let E be any infinite set, T the finite complement topology on E. Prove that T is T_1 but not Hausdorff.

It is clear that T is T_1. Show that if it were Hausdorff then E would have to be a finite set.

Exercise 143. Show that the topology T_d induced by a metric d on a set E is Hausdorff.

Exercise 144. Let E be the set $\mathbf{N} \cup \{\infty\}$ where ∞ is an object which is not a natural number. Let q be the mapping from $E \times E$ to $\mathbf{R}$ defined as follows:

$q(x,x) = 0$ for all x in E ;
$q(x,0) = 1$ for all $x \neq 0$ in E ;
$q(x,\infty) = 1$ for all $x \neq \infty$ in E ;
$q(0,n) = q(\infty,n) = 1/n$ for all positive integers n ;
$q(m,n) = |1/m - 1/n|$ for all positive integers m, n .

Show that q is a quasimetric on E and that the topology induced by q on E is not Hausdorff. (Show that 0 and ∞ have no disjoint neighbourhoods.)

Exercise 145. Let E be the set $\{(x,y) \in \mathbf{R}^2 : x \text{ and } y \text{ are rational}$ numbers and $y \geq 0\}$. Let θ be an irrational number. For each point (x,y) of E and each positive real number ε define the set
$N_\varepsilon(x,y) = \{(x,y)\} \cup \{(t,0) : t \text{ is rational and } |t - x - \theta y| < \varepsilon\}$
$\qquad\qquad \cup \{(t,0) : t \text{ is rational and } |t-x+\theta y| < \varepsilon\}.$
($N_\varepsilon(x,y)$ consists of (x,y) together with two intervals on the rational X-axis centred at the points where the lines through (x,y) with slopes $1/\theta$ and $-1/\theta$ meet the X-axis.) Let $V(x,y)$ be the collection of subsets N of E such that N includes some set $N_\varepsilon(x,y)$. The family of sets $(V(x,y))_{(x,y)\in E}$ satisfies the conditions of Theorem 6 of Chapter 1; so there is a topology T_θ on E such that, for every point (x,y) of E, the collection $V(x,y)$ is the T_θ-neighbourhood filter of (x,y). Prove that the topology T_θ is Hausdorff.

Draw a picture; find the centres of the intervals which form part of $N_\varepsilon(x_1,y_1)$ and $N_\varepsilon(x_2,y_2)$. Choose ε so small that these intervals do not overlap.

Theorem 2 = Exercise 146. Let (E, T) be a topological space. Then the following conditions are equivalent:

(1) T is Hausdorff;

(2) For every point x of E the intersection of the family of T-closed neighbourhoods of x is x;

(3) If a filter F on E converges to a point x then x is the only adherent point of F;

(4) A filter F on E can have at most one limit point.

To prove the implication (1) $\Longrightarrow$ (2) show that, for every point y distinct from x, there is a closed neighbourhood of x not containing y.

The proof that (2) $\Longrightarrow$ (3) follows from the remark that a point which is adherent to a filter converging to x must be in every closed neighbourhood of x.

It is obvious that (3) $\Longrightarrow$ (4).

If (4) holds and T is not Hausdorff then there are distinct points x and y such that every neighbourhood of x meets every neighbourhood of y. This observation allows us to construct a filter which converges both to x and to y.

Exercise 147. Let (E, T) be a topological space, A a subset of E. Prove that if T is T_k then the subspace topology T_A is also T_k $(k = 0, 1, 2)$.

Exercise 148. Let $((E_i, T_i))_{i \in I}$ be a non-empty family of non-empty topological spaces, (E, T) the topological product of this family. Prove that (E, T) is T_k if and only if all the spaces (E_i, T_i) are T_k $(k = 0, 1, 2)$.

Exercise 149. Show that a topological space (E, T) is Hausdorff if and only if the diagonal D of $E \times E$ is $(T \times T)$-closed.

Let P be the complement of D in $E \times E$.

If (E, T) is Hausdorff we show that D is closed by showing that P is open; and we do this by showing that it is a $(T \times T)$-neighbourhood of each of its points (x, y). Notice that if $(x, y) \in P$ then x and y are distinct points of E, so we may apply the Hausdorff condition.

Conversely, if D is closed then P is open. If x and y are distinct points of E then $(x, y) \in P$ and so P is a neighbourhood of (x, y).

Exercise 150. Let (E, T) be any topological space, (E', T') a Hausdorff space; let f and g be (T, T')-continuous mappings from E to E'. Prove that the set $A = \{x \in E : f(x) = g(x)\}$ is T-closed.

If $t \notin A$ then $f(t) \neq g(t)$, so we can apply the Hausdorff condition on E' to obtain disjoint open sets containing these points. Use the continuity of f and g to find neighbourhoods U and V of t whose intersection does not meet A.

Exercise 151. Let (E, T), (E', T') and f be as in Exercise 150. Prove that the graph of f, i.e. the set G of points (x, y) in $E \times E'$ such that $y = f(x)$, is a $(T \times T')$-closed subset of $E \times E'$.

If (x, y) is a point of the complement of the graph (which we want to prove open) then $y \neq f(x)$, so we can apply the Hausdorff condition to obtain disjoint T'-open sets U' and V' containing $f(x)$ and y respectively. The continuity of f gives a T-open subset U of E containing x such that $f^{\rightarrow}(U) \subseteq U'$. Show that $U \times V'$ does not meet G.

Exercise 152. Show that if E is a finite set then the only Hausdorff topology on E is the discrete topology.

This follows easily from Exercise 140. Alternatively we may apply the Hausdorff property to all the pairs $(x_1, x_2), \ldots, (x_1, x_n)$ to obtain open sets $U_2, \ldots, U_n$ containing x_1 but not $x_2, \ldots, x_n$; then consider the intersection of these open sets.

(4) The topology T and the space (E, T) are said to be $\mathbf{T_{2\frac{1}{2}}}$ or **completely Hausdorff** if, for every pair of distinct points x, y of E, there exist T-neighbourhoods V, W of x, y respectively such that $\mathrm{Cl}\, V \cap \mathrm{Cl}\, W = \emptyset$.

Exercise 153. Show that the irrational slope topology of Exercise 145 is Hausdorff but not completely Hausdorff.

Show that the closure of $N_\varepsilon(x, y)$ includes the union of two infinite stripes forming a saltire. Two such saltires must intersect.

Exercise 154. Let $E = \{(x, y) \in \mathbf{R}^2 : y \geq 0\}$. For each point (x, y) of E such that $y > 0$ let $B(x, y)$ be the collection of discs $D((x, y), r)$ with centre (x, y) and radii r where $0 < r \leq y$; for each point $(a, 0)$ of E let $B(a, 0)$ be the collection of sets of the form

$$D((a, 0), r) = \{(a, 0)\} \cup \{(x, y) \in E : y > 0 \text{ and } (x - a)^2 + y^2 < r^2\}$$

where r is a positive real number.

For all points (x, y) of E, let $V(x, y)$ be the collection of subsets N of E such that N includes some set in $B(x, y)$. The family of sets $(V(x, y))_{(x,y) \in E}$ satisfies the conditions of Theorem 6 in Chapter 1; so there is a topology T on E such that, for every point (x, y) of E, the collection $V(x, y)$ is the T-neighbourhood filter of (x, y). The topology T is called the **half-disc topology** on E.

Show that the half-disc topology is completely Hausdorff.

Draw some pictures and show that if p and q are distinct then the closures of the $d/3$-balls round p and q do not meet (where d is the ordinary Euclidean distance between p and q).

(5) The topology T and the space (E, T) are said to be **regular** if, for every point x of E and every T-closed subset A not containing x, there exist disjoint T-open sets U and V such that $x \in U$ and $A \subseteq V$.

The topology T and the space (E, T) are said to be **T₃** if they are both T_1 and regular.

There is a certain inconsistency in the literature over the terminology used in this section (and the following three): some authors use *regular* where we have used T_3 and *vice versa*. We have chosen to follow what appears to be the majority, confirmed in our choice by the result of Exercise 174.

Exercise 155. Show that the half-disc topology on the upper half plane is not regular.

Let $p = (a, 0)$ be a point of the horizontal axis, F the complement in E of $D(p, 1)$. Show that F includes the union of the intervals $(a - 1, a)$ and $(a, a + 1)$ on the axis and that every open set containing p meets this union.

Exercise 156. Show that the digital topology on $\mathbf{Z}$ is not regular.

Exercise 157. Let $E = \{a, b, c\}, T = \{\emptyset, E, \{a\}, \{b, c\}\}$. Prove that T is regular but not T_1.

List the T-closed sets and the pairs (x, F) consisting of a point x and a T-closed set F not containing x; in each case look in T for appropriate non-intersecting open sets.

T is not T_1 since not every singleton is closed.

Exercise 158. Let (E, T) be a topological space. Prove that T is regular if and only if, for every point x of E and every T-open set U containing x, there exists a T-open set U' containing x such that $\mathrm{Cl}\, U' \subseteq U$.

If T is regular and U is a T-open set containing x, apply the regularity property to x and the T-closed set $C_E(U)$.

If the condition holds and F is a T-closed set not containing x, apply the condition with $U = C_E(F)$.

Exercise 159. Show that a topological space (E, T) is regular if and only if each point of E has a fundamental system of T-neighbourhoods consisting of T-closed sets.

Suppose T is regular. To show that each point has a fundamental system of closed T-neighbourhoods, we must find included in each T-neighbourhood of x a closed T-neighbourhood. Each T-neighbourhood V of x includes an open set U containing x; to this open set apply the result of Exercise 158. To prove the converse we apply Exercise 158 in the reverse direction.

Exercise 160. Show that every subspace of a regular space is regular.

(6) The topology T and the space (E, T) are said to be **completely regular** if, for every point x of E and every T-closed subset A of E not containing x, there exists a continuous mapping f from E to the closed interval $[0, 1]$ such that $f(x) = 0$ and $f(t) = 1$ for all points t in A. T and (E, T) are said to be $\mathbf{T_{3\frac{1}{2}}}$ if they are both T_1 and completely regular. A $T_{3\frac{1}{2}}$ space is often called a **Tihonov space**.

As we remarked in Section (5) the terminology in the literature is not entirely consistent: some authors interchange the meanings of *completely regular* and $T_{3\frac{1}{2}}$.

There is an example of a space, called the **Tihonov corkscrew**, which is T_3 but not $T_{3\frac{1}{2}}$. [See Steen and Seebach, *Counterexamples in topology*, page 109.]

Exercise 161. Let $E = \{(x, y) \in \mathbf{R}^2 : y \geq 0\}$. For each point (x, y) of E such that $y > 0$, let $B(x, y)$ be the collection of sets of the form $E \cap V((x, y), \varepsilon)$ for all positive real numbers ε ; for each point $(x, 0)$ of E, let $B(x, 0)$ be the collection of sets of the form $V((x, \varepsilon), \varepsilon) \cup \{(x, 0)\}$ for all positive real numbers ε ; let $B = \bigcup_{(x,y) \in E} B(x, y)$. Then B is base for a topology T on E, called the **Nemytskii tangent disc topology**. Show that T is $T_{3\frac{1}{2}}$.

To show that the Nemytskii topology is T_1, we take distinct points p and q and consider the cases where (1) p and q are not on the horizontal axis; (2) p and q are both on the horizontal axis; (3) one of p, q is on the horizontal axis and the other is not. In each case it is easy to find neighbourhoods of p, q satisfying the T_1 condition by drawing pictures.

Next let F be a closed subset of E and p a point not in F, so that there is a T-neighbourhood V of p included in $C_E(F)$. Distinguish the cases where (1) p is not on the horizontal axis and (2) p is on the axis and draw pictures to construct continuous functions from E to $[0, 1]$ which take the value 0 at p and 1 at all points of F.

Exercise 162. Show that every completely regular space is regular.

To show that a completely regular space (E, T) is regular, let a be a point of E and F a T-closed subset not containing a. There is a continuous mapping $f : E \to [0, 1]$ for which we have $f(a) = 0$ and $f^\to(F) = \{1\}$. Take non-overlapping open subsets of $[0, 1]$ containing 0 and 1 and look at their inverse images under f.

Exercise 163. Show that the topological product of a family of completely regular spaces is completely regular.

Let a be a point of the product space, F a closed subset not containing a. Then the complement of F is open and so includes a product of open subsets U_i of the factor spaces, only finitely many of which (say for i in some finite set J) are not the whole space. Use the complete regularity of the factor spaces E_i with i in J to produce continuous functions $f_i : E_i \rightarrow [0,1]$ which take the value 0 at the i-th projection of a and 1 at all points of the complement of U_i. Use these functions f_i to construct a continuous function f from the product space to $[0,1]$ such that $f(a) = 0$ and $f^{\rightarrow}(F) = \{1\}$.

Theorem 3 = Exercise 164 (Tihonov's Embedding Theorem). A topological space is a Tihonov ($T_{3\frac{1}{2}}$) space if and only if it is homeomorphic to a subspace of the product of a family of spaces all equal to [0,1] with its usual metric topology.

The *if* part is straightforward.

To prove the *only if* part, let (E,T) be a completely regular space. Let F be the set of all continuous mappings from E to $[0,1]$; for each f in F let $I_f = [0,1]$ and let $P = \prod_{f \in F} I_f$. Define a mapping g from E to P by setting $g(x) = (f(x))_{f \in F}$ for all x in E. Prove first (using the T_1 property) that g is injective (so that we have an inverse mapping g^{-1} from $g^{\rightarrow}(E)$ to E); then that g is continuous; then finally that g^{-1} is continuous.

(7) The topology T and the topological space (E,T) are said to be **normal** if, for every pair of disjoint T-closed subsets A and B of E, there exist disjoint T-open subsets U and V including A and B respectively. T and (E,T) are said to be **T_4** if they are both T_1 and normal.

Here again there is some confusion over terminology: some authors use T_4 where we have used *normal* and *normal* where we have used T_4.

It can be shown that the Nemytskii space is $T_{3\frac{1}{2}}$ but not T_4. [See Steen and Seebach, *Counterexamples in Topology*, page 101.]

Let ω_0 be the first infinite ordinal, ω_1 the first uncountable ordinal, $\Omega_0 = [0, \omega_0]$, $\Omega_1 = [0, \omega_1]$ with the order topology in each case. Then $TP = \Omega_0 \times \Omega_1$ equipped with the product topology is called the **Tihonov plank**; it can be shown to be a T_4 space.

Exercise 165. Show that the digital topology on **Z** is not normal.

Exercise 166. Let (E,T) be a topological space. Show that (E,T)

is normal if and only if, for every T-open set U and every T-closed subset A of U, there is a T-open set U' such that $A \subseteq U'$ and $\operatorname{Cl} U' \subseteq U$.

Theorem 4 = Exercise 167 (Urysohn's Lemma). Let (E, T) be a normal topological space. If A and B are disjoint T-closed subsets of E there exists a continuous mapping f from E to $[0,1]$ such that $f^{\rightarrow}(A) = \{0\}$ and $f^{\rightarrow}(B) = \{1\}$.

Let $U_1 = C_E(B)$; then U_1 is a T-open subset including A. According to Exercise 166 there is a T-open set U_0 such that $A \subseteq U_0$ and $\operatorname{Cl} U_0 \subseteq U_1$. By Exercise 166 again, applied to $\operatorname{Cl} U_0$ and U_1, there is a T-open set $U_{\frac{1}{2}}$ such that $\operatorname{Cl} U_0 \subseteq U_{\frac{1}{2}}$ and $\operatorname{Cl} U_{\frac{1}{2}} \subseteq U_1$.

Show how to use Exercise 166 to construct inductively for all dyadic rationals r (i.e. rational numbers with denominator a power of 2) in $[0, 1]$ open sets U_r in such a way that for all such dyadic rationals r_1, r_2 with $r_1 < r_2$ we have $\operatorname{Cl} U_{r_1} \subseteq U_{r_2}$.

Now define $f : E \to [0, 1]$ by setting

$$f(x) = \begin{cases} 1 & \text{if } x \in B \\ \inf \{ r \in [0, 1] : x \in U_r \} & \text{if } x \notin B. \end{cases}$$

Certainly $f^{\rightarrow}(A) = \{0\}$ and $f^{\rightarrow}(B) = \{1\}$.

We have to show that f is continuous. The usual topology on $[0, 1]$ is generated by the collection of subsets of the two types $[0, s)$ and $(t, 1]$; so we need only show that the inverse images under f of such sets are T-open. To do this we prove first that $f^{\leftarrow}[0, s) = \bigcup_{r \in K} U_r$ where K is the set of dyadic rationals r in $[0, s)$ and then that $f^{\leftarrow}(t, 1] = \bigcup_{r \in L} (C_E(\operatorname{Cl} U_r))$ where L is the set of dyadic rationals in $(t, 1]$.

Theorem 5 = Exercise 168 (Tietze's Extension Theorem). Let (E, T) be a normal topological space, F a T-closed subset of E. Then every continuous mapping from F to a bounded closed interval I of $\mathbf{R}$ can be extended to a continuous mapping from E to I.

Take the bounded closed interval I to be $[-1, 1]$. Show that $A = f^{\leftarrow}[-1, -\frac{1}{3}]$ and $B = f^{\leftarrow}[\frac{1}{3}, 1]$ are disjoint T-closed subsets of E. Modify Urysohn's Lemma to get a continuous mapping g from E to $[-\frac{1}{3}, \frac{1}{3}]$ such that $g^{\rightarrow}(A) = \{-\frac{1}{3}\}$ and $g^{\rightarrow}(B) = \{\frac{1}{3}\}$. Set $f_0 = f$, $g_0 = g$, and let f_1 be the restriction of $f_0 - g_0$ to F. Prove that $f_1^{\rightarrow}(F) \subseteq [-\frac{2}{3}, \frac{2}{3}]$.

Repeating this performance define sequences of mappings

$f_n : F \rightarrow [-(\frac{2}{3})^n, (\frac{2}{3})^n]$ and $g_n : E \rightarrow [-\frac{1}{3}(\frac{2}{3})^n, \frac{1}{3}(\frac{2}{3})^n]$ with f_{n+1} the restriction of $f_n - g_n$ to F.

Use Cauchy's criterion for uniform convergence to show that the series $\sum g_n$ is uniformly convergent on E; it follows that the sum function g is continuous. Prove finally that $g(x) = f(x)$ for all points x of F.

Exercise 169. Let (E, T) be a topological space. If $E = F_1 \cup F_2$ where F_1 and F_2 are T-closed subsets of E, each of which is normal in its subspace topology, prove that (E, T) is normal.

Let A and B be disjoint T-closed subsets of E. Then $A \cap F_1$ and $B \cap F_1$ are disjoint T_{F_1}-closed subsets of F_1. Since T_{F_1} is normal there are disjoint T_{F_1}-open sets including these subsets: these are intersections of F_1 with T-open subsets of E, say U_1 and V_1. Let $U_1' = U_1 \cup C_E(F_1)$ and $V_1' = V_1 \cup C_E(F_1)$; these are T-open subsets of E including A and B respectively. Define U_2' and V_2' similarly using the normality of T_{F_2}. Then show that $U = U_1' \cap U_2'$ and $V = V_1' \cap V_2'$ are disjoint T-open sets which include A and B respectively.

Exercise 170. Let (E, T) be a normal space, F a T-closed subset of E; and let f be a continuous mapping from the subset F to $I^2 = \{(x, y) \in \mathbf{R}^2 : 0 \leq x, y \leq 1\}$. Prove that there exists a continuous mapping g from E to I^2 such that the restriction of g to F is f.

This is a simple application of Tietze's Extension Theorem.

(8) Let A and B be subsets of E; then A and B are said to be **separated** if $A \cap \mathrm{Cl}\,B = \emptyset = B \cap \mathrm{Cl}\,A$. (E, T) and T are said to be **completely normal** if, for every pair of separated subsets A, B of E, there exist disjoint T-open subsets U and V such that $A \subseteq U$ and $B \subseteq V$. (E, T) and T are said to be **T_5** if they are both completely normal and T_1. Once again there is disagreement over terminology, with some books using T_5 for our *completely normal* and *vice versa*.

It can be shown that the Tihonov plank is not completely normal.

Exercise 171. Let (E, d) be a metric space. Then the topology T_d induced by d is completely normal.

Let A and B be separated subsets of E. Each point a of A is in the complement of the closure of B, which is an open set; so there is a ball with centre a entirely included in the complement, say $V_d(a, r(a))$. Let $U = \bigcup_{a \in A} V_d(a, \frac{1}{2}r(a))$; U is an open set including A. Define in the same way an open set V including B and show that U and V are disjoint.

Theorem 6 = Exercise 172. A topological space (E, T) is completely normal if and only if every subspace is normal.

(1) Suppose (E, T) is completely normal; we can actually prove that every subspace is *completely* normal. To do this we show, using part of Exercise 77, that if A and B are separated subsets relative to the subspace topology of a subset F of E then they are actually separated subsets of E relative to T.

(2) Suppose now that all the subspaces of (E, T) are normal. Let A and B be separated subsets of E. Apply the normality of the subspace $X = C_E(\mathrm{Cl}_T(A) \cap \mathrm{Cl}_T(B))$ to its disjoint closed subsets $X \cap \mathrm{Cl}_T(A)$ and $X \cap \mathrm{Cl}_T(B)$.

Exercise 173. Let $E = \mathbf{R}$. Let T be the topology on E consisting of all sets of the form $U \cup V$ where U is an ordinary open set and V is a subset of the irrationals. Prove that (E, T) is T_4. (This space (E, T) is called the **scattered line**.)

Let T^* be the ordinary Euclidean topology on E. Use the fact that $T^* \subseteq T$ to deduce that T is T_1.

Now let A and B be disjoint T-closed subsets of E. Let A_1 and B_1 be their intersections with $\mathbf{Q}$ (the set of rational numbers) and show that these are T^*-separated subsets of E. Since T^* is completely normal (by Exercise 171) there are disjoint T^*-open subsets U_1 and V_1 of E including A_1 and B_1 respectively. Let $U = U_1 \cup (A \cap C_E(\mathbf{Q}))$, $V = V_1 \cup (B \cap C_E(\mathbf{Q}))$; show that U and V are T-open disjoint sets which include A and B respectively.

Exercise 174. Let (E, T) be a topological space. Let i, j be members of the set $\{0, 1, 2, 2\frac{1}{2}, 3, 3\frac{1}{2}, 4, 5\}$. Prove that if T is a T_j topology and $i < j$, then T is also a T_i topology.

Chapter 6

COMPACTNESS

Let E be a set, A a subset of E. A **cover** of A is a family of subsets of E whose union includes A; a **subcover** of a cover of A is a subfamily which is also a cover of A. If (E, T) is a topological space and A is a subset of E then an **open cover** of A is a cover in which all the subsets are T-open.

Let (E, T) be a topological space. A subset A of E is said to be **compact** if every T-open cover of A has a finite subcover. (E, T) is said to be **locally compact** if every point of E has a compact T-neighbourhood.

A family $(F_i)_{i \in I}$ of subsets of a set E is said to have the **finite intersection property** if for every finite subset J of I the intersection $\bigcap_{i \in J} F_i$ is non-empty.

Exercise 175. Let E be an infinite set, T the finite complement topology on E. Prove that (E, T) is compact.

After you have used one set of the cover not much of E is left uncovered!

Exercise 176. Prove that a discrete topological space is compact if and only if it is finite.

In fact every finite space is compact. If E is an infinite discrete space it is easy to exhibit an open cover with no finite subcover.

Theorem 1 = Exercise 177. Let (E, T) be a topological space, A a subset of E. Then A is compact if and only if, for every family $(F_i)_{i \in I}$ of T_A-closed subsets of A with the finite intersection property, we have $\bigcap_{i \in I} F_i \neq \emptyset$.

This is an easy exercise on the use of the De Morgan properties that $C_E(\bigcap_{i \in I} X_i) = \bigcup_{i \in I} C_E(X_i)$ and $C_E(\bigcup_{i \in I} X_i) = \bigcap_{i \in I} C_E(X_i)$.

Theorem 2 = Exercise 178. Let (E, T) be a topological space. Then the following conditions are equivalent:
(1) (E, T) is compact;
(2) Every filter on E has at least one adherent point;
(3) Every ultrafilter on E converges.

To prove that (1) $\Longrightarrow$ (2) notice that if F is a filter on E then the family $(\mathrm{Cl}\, X)_{X \in F}$ of closed sets has the finite intersection property.

For the implication (2) $\Longrightarrow$ (3) recall that every adherent point of an ultrafilter is a limit point.

To prove that (3) $\Longrightarrow$ (1) consider a family of closed sets with the finite intersection property. This family generates a filter, which is included in an ultrafilter, for which there is a limit point p. Show that p is in each of the closed sets of the family.

Theorem 3 = Exercise 179. Let (E, T) be a Hausdorff space. Then every compact subset of E is T-closed.

Let K be a compact subset of E. Let a be a point not in K; applying the Hausdorff property to a and each point x of K, we obtain disjoint open sets containing a and x respectively. This gives an open cover of K which has a finite subcover. Use the open sets containing a which correspond to the open sets of the subcover to construct a neighbourhood of a which does not meet K.

Exercise 180. Prove that the union of each finite family of compact subsets of a topological space is compact.

An open cover of the union is of course a cover of each of the compact sets.

Theorem 4 = Exercise 181. Let (E, T) be a compact topological space. Every T-closed subset of E is compact.

An open cover of the closed subset F together with the complement

$C_E(F)$ is an open cover of E.

Exercise 182. Prove that the intersection of each non-empty family of compact subsets of a Hausdorff space is compact.

Use Exercises 179 and 181.

Theorem 5 = Exercise 183. Every compact Hausdorff space is normal.

Adapt the approach of Exercise 179.

Theorem 6 = Exercise 184. Let (E, T) be a compact topological space, (E', T') any topological space and f a (T, T')-continuous mapping from E to E'. Then $f^{\rightarrow}(E)$ is compact.

This is clear if we use the "inverse image of open sets" definition of continuity.

Theorem 7 = Exercise 185. Let (E, T) be a compact space, (E', T') a Hausdorff space and f a (T, T')-continuous bijection from E onto E'. Then f is a (T, T')-homeomorphism.

All that remains to show is that f^{-1} is (T', T)-continuous. So take a T-closed subset F of E and try to show, using Exercises 181, 184 and 179 that $(f^{-1})^{\leftarrow}(F) = f^{\rightarrow}(F)$ is T'-closed.

Theorem 8 = Exercise 186. (Tihonov's Theorem). Let $(E_i, T_i)_{i \in I}$ be a family of topological spaces, (E, T) its topological product. If all the spaces (E_i, T_i) are compact then so is (E, T). If all the sets E_i are non-empty and (E, T) is compact then all the spaces (E_i, T_i) are compact.

(1) Suppose all the factor spaces are compact; let U be an ultrafilter on the product space. For each index i the collection of the i-th projections of all the sets in U is base for an ultrafilter U_i on E_i (see Exercise 108). Each of these ultrafilters U_i converges, to a_i say. Prove that U converges to the point $a = (a_i)$ of the product.

(2) If all the sets E_i are non-empty and the product (E, T) is compact, then the projections π_i are (T, T_i)-continuous surjections.

A topological space (E, T) is said to be **countably compact** if every countable open cover of E has a finite subcover.

Theorem 9 = Exercise 187. Let (E, T) be a topological space. The following conditions are equivalent:

(1) (E, T) is countably compact;

(2) Every countable family of T-closed subsets of E with the finite intersection property has non-empty intersection;

(3) Every countably infinite subset of E has an ω-accumulation point;

(4) Every sequence in E has an adherent point.

The equivalence (1) $\Longleftrightarrow$ (2) is proved using the same approach as Exercise 177.

To prove (3) $\Longrightarrow$ (4) let $\sigma : \mathbf{N} \to E$ be a sequence of points in E; let $A = \sigma^{\to}(\mathbf{N})$. Distinguish the cases where A is finite and A is infinite.

For the implication (4) $\Longrightarrow$ (3) notice that every countably infinite subset of E gives rise to a sequence of points in E (by the definition of countability).

To prove that (3) $\Longrightarrow$ (1) suppose (3) holds. Let (U_n) be a countable open cover of E. Without loss of generality we may assume that all the sets U_n are distinct and that none is included in the union of the sets U_i with $i < n$. If this cover has no finite subcover construct an infinite set X with no ω-accumulation point.

Finally, to prove that (1) $\Longrightarrow$ (3) suppose there is a countably infinite subset S of E with no ω-accumulation point. Let Z be the set of all finite subsets of S; Z is countable. Since S has no ω-accumulation points it follows that every point x of E has an open neighbourhood U_x such that $S \cap U_x$ is finite. For each (finite) set F in Z let U_F be the union of the sets U_x which meet S in F. Show that $(U_F)_{F \in Z}$ is a countable open cover of E with no finite subcover.

Exercise 188. Every compact space is countably compact.

Exercise 189. Prove that if (E, T) is a compact space and (E', T') is a countably compact space then $(E \times E', T \times T')$ is countably compact.

The arguments here are a bit delicate. Let $(U_n)_{n \in \mathbf{N}}$ be a countable open cover of $E \times E'$. Form an increasing sequence (G_n) of open sets of $E \times E'$ by defining $G_n = \bigcup_{k \leq n} U_k$ for all natural numbers n. For each n let H_n be the set of points y in E' for which there is a T'-neighbourhood V' of y' such that $E \times V' \subseteq G_n$. Show that all these sets H_n are T'-open and that (H_n) is an increasing sequence.

Since E is compact, so is every subset of $E \times E'$ of the form $E \times \{b\}$. Using this observation we can show that (H_n) is a (countable) open cover of E'; so it has a finite subcover. Since (H_n) is increasing this means that one of the sets H_n, say H_p must be equal to E'. From this we can deduce that $E \times E' = U_1 \cup ... \cup U_p$, i.e. that (U_n) has a finite subcover.

A topological space (E, T) is said to be **sequentially compact** if every sequence in E has a convergent subsequence.

Exercise 190. Let (E, T), (E', T') be topological spaces, f a (T, T')-continuous mapping from E to E'. Prove that if (E, T) is either countably compact or sequentially compact, then so is $f^{\rightarrow}(E)$.

For the case of countable compactness use the approach of Exercise 184.

For the case where (E, T) is sequentially compact let σ be a sequence in $f^{\rightarrow}(E)$; define a sequence ν in E by setting $\nu(n) = x_n$ where x_n is a point of E such that $f(x_n) = \sigma(n)$. Construct a convergent subsequence of σ from a convergent subsequence of ν.

Exercise 191. Let ω_1 be the first uncountable ordinal with the order topology. Show that ω_1 is sequentially compact but not compact.

We can show that, for every ordinal number γ, its successor $\Gamma = \gamma \cup \{\gamma\}$ equipped with the order topology is compact. To do this let $(U_i)_{i \in I}$ be an open cover of Γ and let S be the subset of Γ consisting of all points y such that the interval $[0, y)$ can be covered by finitely many of the sets U_i. Show that S has a least upper bound, α say, and prove that α must in fact be γ. Deduce that Γ is compact. In particular Ω_1 is compact and hence countably compact.

To show that ω_1 is countably compact, we show that every countably infinite subset of ω_1 has an ω-accumulation point in ω_1. Each such subset certainly has an ω-accumulation point in Ω_1; we show that this cannot be ω_1 and hence must be a member of ω_1. (For if A is a countable subset of ω_1 let α be the least upper bound (which is actually the union) of A; then α is countable, hence not equal to ω_1 and the neighbourhood $(\alpha, \omega_1]$ of ω_1 does not meet A at all.)

Exercise 192. Let $I = [0, 1]$ and let P be the topological product of the family $(X_i)_{i \in I}$ where for each index i in I we have $X_i = I$. Show that P is compact but not sequentially compact.

(1) I is compact by the Heine-Borel Theorem of real analysis; hence P is compact by Tihonov's Theorem.

(2) To show that P is not sequentially compact, prove that if $f : I \to I$ is any point of P, then a sequence (f_n) of points of P converges to f if and only if the sequence $(f_n(x))$ converges to $f(x)$ for all points x of I. Define the sequence (f_n) in P by setting $f_n(x) =$ the n-th digit in the binary expansion of x. Show that this has no convergent subsequence.

Exercise 193. Show that if P is the product space defined in Exercise 192, then $\omega_1 \times P$ is countably compact but not compact and not sequentially compact.

Use Exercise 192 (1) for P, Exercise 191 for ω_1 and Exercise 189 for the product. If $\omega_1 \times P$ were compact then ω_1 would also be compact; if $\omega_1 \times P$ were sequentially compact then P would be as well.

Exercise 194. Let E be an infinite set; let p be a point of E. Show that the Fort topology on E (see Exercise 13) is both compact and sequentially compact.

To show that T is compact consider an open cover and look first at the set of the cover which contains p.

To show that T is sequentially compact let σ be a sequence in E and distinguish the cases where $\sigma^{\to}(\mathbf{N})$ is finite and $\sigma^{\to}(\mathbf{N})$ is infinite; in the second case show that there is a subsequence converging to p.

Theorem 10. Let (E, d) be a metric space, T the topology induced on E by the metric d. Then the following conditions are equivalent:

(1) (E, T) is compact;

(2) (E, T) is countably compact;

(3) (E, T) is sequentially compact.

For the proof of this theorem, which seems to lie just on the analysis side of the dividing line between general topology and real analysis, we refer the reader to Section 24 of G. F. Simmons's book *Introduction to topology and modern analysis*.

Theorem 11 = Exercise 195 (Alexandrov's Theorem). Let (E, T) be a locally compact Hausdorff space. Then there exists a compact Hausdorff space (E_1, T_1), a point ∞ of E_1 and a homeomorphism i from E onto the complement of $\{\infty\}$ in E_1, equipped with its subspace topology.

Let $E_1 = E \cup \{\infty\}$ where ∞ is an object not in E; let $T_1 = T \cup T_0$ where T_0 is the collection of all subsets of E_1 of the form $\{\infty\} \cup C_E(K)$ where K is a compact subset of E.

Verify that T_1 is a topology on E_1 (there are various cases to consider when proving that unions and finite intersections of sets in T_1 belong to T_1).

Let i be the natural injection of E into E_1; it is a homeomorphism from E onto $i^{\rightarrow}(E)$.

To show that T_1 is Hausdorff, let a and b be distinct points of E_1. Distinguish the cases where a and b are both in E and where one of them is the added point ∞; in the second case use the local compactness of (E, T).

To show that T_1 is compact consider an open cover of E_1. One of the sets of the cover must be of the form $\{\infty\} \cup C_E(K)$ where K is a compact subset of E. The intersections with E of the remaining sets of the cover form a cover of K.

Let (E, T) be a completely regular space; let $C^*(E)$ be the set of continuous mappings from E to $I = [0, 1]$ and for each f in $C^*(E)$ let $I_f = I$. Let $P_E = \prod_{f \in C^*(E)} I_f$; P_E is compact since each of its factors is compact. Let e be the mapping from E to P_E such that $e(x)_f = f(x)$ for all points x in E and all mappings f in $C^*(E)$. According to the Tihonov Embedding Theorem e is a homeomorphism from E onto $e^{\rightarrow}(E)$. We define βE to be the closure of $e^{\rightarrow}(E)$; since βE is a closed subset of the compact space P_E we see that βE is compact. Then the ordered pair $(\beta E, e)$ is called the **Stone-Čech compactification** of (E, T).

Theorem 12 = Exercise 196. Let (E, T) be a completely regular space, (E', T') a compact Hausdorff space and f a continuous mapping from E to E'. Then there exists a continuous mapping βf from βE to E' such that $(\beta f) \circ e = f$.

Since (E', T') is compact Hausdorff it is normal and hence completely regular. Define $P_{E'}$ and e' by analogy with P_E and e. For each mapping f' in $C^*(E')$ the projection mapping $\pi_{f' \circ f} : P_E \to I$ is continuous; show that there is a continuous mapping H from P_E to $P_{E'}$ such that $H \circ e = e' \circ f$. Prove that $(e')^{\rightarrow}(E')$ is a closed subset of $P_{E'}$ and deduce that $H^{\rightarrow}(\beta E) \subseteq (e')^{\rightarrow}(E')$. Finally show that $\beta f = (e')^{-1} \circ (H \mid \beta E)$ is the required continuous mapping from βE to E'.

Chapter 7

CONNECTEDNESS

Let (E,T) be a topological space. We recall that two subsets A and B are said to be **separated** if $A \cap \mathrm{Cl}_T(B) = \emptyset = B \cap \mathrm{Cl}_T(A)$.

Exercise 197. Prove that if A and B are disjoint closed subsets of E then they are separated. Prove that if A and B are disjoint open subsets of E then they are separated.

A topological space (E,T) is said to be **connected** if the only subsets of E which are both T-open and T-closed are E and $\emptyset$. A space which is not connected is said to be **disconnected**.

Theorem 1 = Exercise 198. Let (E,T) be a topological space. The following conditions are equivalent:

(1) (E,T) is disconnected;

(2) E is the union of two non-empty separated subsets;

(3) E is the union of two disjoint non-empty T-closed subsets;

(4) E is the union of two disjoint non-empty T-open subsets.

Let (E,T) be a topological space, A a subset of E. Then A is said to be **connected** or **disconnected** according as the space (A, T_A) is connected or disconnected.

When we are trying to prove that a topological space is connected we often proceed by contradiction, supposing that the space is disconnected and then using any of the equivalent conditions of Theorem 1.

Exercise 199. Let (E, T) be a topological space, A a subset of E. Prove that A is disconnected if and only if it can be expressed as the union of two non-empty separated subsets of E.

If A is the union of two disjoint non-empty T_A-closed subsets, show that these subsets are separated subsets of E. If A is the union of two non-empty subsets of E separated relative to the topology T show that these subsets are separated relative to the subspace topology T_A.

Theorem 2 = Exercise 200. Let (E, T) be a topological space; let A be a connected subset of E. Then every set B such that $A \subseteq B \subseteq \mathrm{Cl}_T(A)$ is connected.

Suppose B is disconnected; say $B = X \cup Y$ where X, Y are non-empty separated subsets of E; examine $A \cap X$, $A \cap Y$.

Theorem 3 = Exercise 201. Let (E, T) be a topological space, $(A_i)_{i \in I}$ a family of connected subsets of E. Show that if $\bigcap_{i \in I} A_i$ is non-empty then $A = \bigcup_{i \in I} A_i$ is connected.

Suppose A is disconnected; say $A = X \cup Y$ where X, Y are non-empty separated subsets of E. For each index i in I examine $A_i \cap X$, $A_i \cap Y$.

Exercise 202. Let (E, T) be a topological space; let A and B be connected subsets of E such that $A \cap \mathrm{Cl}_T(B) \neq \emptyset$. Prove that $A \cup B$ is connected.

Suppose $A \cup B = X \cup Y$ where X, Y are non-empty separated subsets of E. Examine $A \cap X$, $A \cap Y$ and $B \cap X$, $B \cap Y$.

Exercise 203. Let (E, T) be a topological space, $(A_n)_{n \in \mathbf{N}}$ a sequence of connected subsets of E such that $A_n \cap A_{n+1} \neq \emptyset$ for every natural number n. Prove that $\bigcup_{n \in \mathbf{N}} A_n$ is connected.

Consider the sets $B_n = A_0 \cup A_1 \cup \ldots \cup A_n$; show that these are connected, with non-empty intersection. Apply Exercise 201.

Exercise 204. Show that $(\mathbf{Z}, T)$ is connected, where T is the digital topology on $\mathbf{Z}$.

Prove that if $\mathbf{Z} = A \cup B$ where A and B are separated then, if an integer k is in A, so also are $k+1$ and $k-1$.

Exercise 205. Prove that a subset of $\mathbf{R}$ with its usual topology is connected if and only if it is an interval.

It is easy to show that if a subset is not an interval then, it is not connected.

To show that an interval E is connected, we proceed by contradiction, i.e. we suppose that we have a disconnection $E = A \cup B$. Take points a and b from A and B respectively; suppose $a < b$. Let $c = \sup([a, b] \cap A)$ and show that $c \in A \cap B$.

Theorem 4 = Exercise 206. Let (E, T), (E', T') be topological spaces, f a (T, T')- continuous mapping from E to E'. If A is a connected subset of E, then $f^{\rightarrow}(A)$ is a connected subset of E'.

Suppose $A' = f^{\rightarrow}(A)$ is disconnected. Then A' is the union of two disjoint non-empty sets open in the subspace topology on A', say $A' \cap X'$ and $A' \cap Y'$ where X', Y' are T'-open subsets of E'. Examine $A \cap f^{\leftarrow}(X')$, $A \cap f^{\leftarrow}(Y')$.

Exercise 207. Let f be a continuous mapping from an interval E of $\mathbf{R}$ to $\mathbf{R}$. If a and b are points of E and k is a real number such that $f(a) < k < f(b)$ prove that there is a point c of E such that $f(c) = k$.

Exercise 208. Let (E, T) be a topological space, (E', T') a discrete space with two points. Prove that (E, T) is disconnected if and only if there exists a (T, T')-continuous surjection from E onto E'.

(1) Suppose E is disconnected; say $E = U \cup V$ where U, V are disjoint non-empty T-open subsets. Define the mapping $f : E \to E'$ by

$$f(x) = \begin{cases} 0 & \text{if } x \in U \\ 1 & \text{if } x \in V. \end{cases}$$

(2) If there is a continuous surjection f from E onto E' examine $f^{\leftarrow}\{0\}$, $f^{\leftarrow}\{1\}$.

Theorem 5 = Exercise 209. Let $((E_i, T_i))_{i \in I}$ be a family of topological spaces, (E, T) its topological product. If all the spaces in the family are connected then the product is connected. If all the sets E_i are non-empty and the product is connected, then all the spaces (E_i, T_i) are connected.

(1) Think first of the product (E, T) of *two* connected spaces (E_1, T_1), (E_2, T_2). Suppose the product is disconnected. Then $E = U \cup V$ where U, V are disjoint non-empty T-open sets. Let u and v be points of U and V respectively. Let $A_1 = \{(x, y) \in E : y = \pi_2(u)\}$ and $A_2 = \{(x, y) \in E : x = \pi_1(v)\}$. Then (A_1, T_{A_1}) and (A_2, T_{A_2}) are homeomorphic to (E_1, T_1) and (E_2, T_2) respectively and hence are connected; $A_1 \cap A_2$ is non-empty since it contains $(\pi_1(v), \pi_2(u))$. So $A = A_1 \cup A_2$ is connected. Examine $A \cap U$, $A \cap V$ to reach a contradiction.

The proof for the case of the product of an arbitrary family of connected spaces is a bit more complicated, but it is based on the same idea.

(2) If all the sets E_i are non-empty and the product (E, T) is connected, then the projections π_i are (T, T_i)-continuous surjections.

Let (E, T) be a topological space, x a point of E. The union of all the connected subsets of E which contain x is clearly connected, and is the largest connected subset of E containing x; we call it the **connected component** of x; clearly x belongs to its own connected component.

A topological space (E, T) is said to be **totally disconnected** if, for every point x of E, the connected component of x is $\{x\}$.

Exercise 210. Let (E, T) and (E', T') be connected topological spaces. Let A and A' be proper subsets of E and E' respectively. Prove that the complement of $(A \times A')$ in $(E \times E')$ is connected.

Exercise 211. Show that every discrete topological space is totally disconnected.

Exercise 212. Show that the set of rational numbers together with its topology as a subspace of the real line is totally disconnected but not discrete.

If x is a rational number then $\{x\}$ is connected; but if A is a subset of $\mathbf{Q}$ which properly includes $\{x\}$ and y is a point of A distinct from x then there is an irrational number z between x and y and A is disconnected by its intersections with $(-\infty, z)$ and (z, ∞).

Theorem 6 = Exercise 213. Let (E, T) be a topological space. For every point x of E, the connected component of x is T-closed. The relation R on E defined by setting $(x, y) \in R$ if and only if y belongs to the connected component of x is an equivalence relation on E and the quotient space $(E/R, T/R)$ is totally disconnected.

To prove the first assertion notice that if $K(x)$ is the connected component of x then its closure is connected.

The proof that R is an equivalence relation is not hard if we keep remembering the definition of the component of a point.

Let C be an element of E/R and K its connected component. If K contains more than one point, then $\eta^{\leftarrow}(K)$ includes more than one connected component of E and hence is disconnected, say by its intersections A_1 and B_1 with two closed subsets of E. Show that A_1 and B_1 are closed and that they are unions of complete R-classes. Deduce that $\eta^{\rightarrow}(A_1)$, $\eta^{\rightarrow}(B_1)$ disconnect K.

A topological space (E, T) is said to be **locally connected** if, for every point x of E and every T-neighbourhood V of x, there exists a connected T-neighbourhood of x included in V.

Exercise 214. Show that every discrete space with more than one point is locally connected but not connected.

Exercise 215. Consider the subset $S = A \cup B$ of $\mathbf{R}^2$ where $A = \{(x, y) \in \mathbf{R}^2 : x > 0 \text{ and } y = \sin(1/x)\}$ and $B = \{(0, 0)\}$. Prove that S is connected but not locally connected (as a subspace of $\mathbf{R}^2$ with its usual topology).

A is connected, being the image of a connected set $\mathbf{R}^+$ under a continuous mapping; $B \cap \mathrm{Cl}A \neq \emptyset$.

Theorem 7 = Exercise 216. A topological space (E, T) is locally connected if and only if the connected components of every T-open subset of E are T-open.

Let (E,T) be locally connected, U an open subset of E, K a component of U, t a point of K. Then U is a neighbourhood of t and so there is a connected neighbourhood of t which must be included in K.

Suppose conversely that every component of every open subset is open. Let V be a neighbourhood of a point x. Then there is an open set U included in V. Look at the component of x in U.

Exercise 217. Prove that every quotient space of a locally connected space is locally connected.

If (E,T) is a topological space and R is an equivalence relation on E let $\bar{K}$ be a connected component of a T/R-open subset $\bar{U}$ of E/R. Let $U = \eta^{\leftarrow}(\bar{U})$. Show that $\eta^{\leftarrow}(\bar{K})$ is the union of the components (in U) of all its points, hence T-open.

Exercise 218. Let $((E_i, T_i))_{i \in I}$ be a family of topological spaces, (E,T) the topological product of the family. Prove that if all the spaces (E_i, T_i) are locally connected and all but a finite number of them are connected then (E,T) is locally connected.

Let $J = \{i \in I : (E_i, T_i) \text{ is not connected}\}$.

Let x be any point of E, V a T-neighbourhood of x. To produce a connected T-neighbourhood of x included in V, recall that V includes a product $\prod_{i \in I} U_i$ where each U_i is T_i-open and $U_i = E_i$ for all indices i not in a certain finite subset K of I. Use the local connectedness of the spaces (E_i, T_i) with i in $J \cup K$.

Part II

ANSWERS

Chapter 8

ANSWERS FOR CHAPTER 1

1. (a) $d(x, y) = |y - x| \geq 0$ for all x, y in $\mathbf{R}$.

$d(x, y) = 0 \iff |y - x| = 0 \iff y - x = 0 \iff y = x$.

(b) $d(x, y) = |y - x| = |x - y| = d(y, x)$ for all x, y in $\mathbf{R}$.

(c) For all x, y, z in $\mathbf{R}$ we have

$d(x, z) = |z - x| = |(z - y) + (y - x)| \leq |z - y| + |y - x| = d(x, y) + d(y, z)$.

So d is a metric on $\mathbf{R}$.

2. (a) $d(x, y) = \sqrt{(\sum_{i=1}^{n}(y_i - x_i)^2)} \geq 0$ for all x, y in $\mathbf{R}^n$.

$d(x, y) = 0 \iff \sum_{i=1}^{n}(y_i - x_i)^2 = 0 \iff y_i = x_i$ for $i = 1, \ldots, n \iff y = x$.

(b) $d(x, y) = \sqrt{(\sum_{i=1}^{n}(y_i - x_i)^2)} = \sqrt{(\sum_{i=1}^{n}(x_i - y_i)^2)} = d(y, x)$ for all x, y in $\mathbf{R}^n$.

(c) For all x, y, z in $\mathbf{R}^n$ we have

$(d(x, y) + d(y, z))^2$

$= \sum_{i=1}^{n}(y_i - x_i)^2 + \sum_{i=1}^{n}(z_i - y_i)^2$
$\qquad + 2((\sum_{i=1}^{n}(y_i - x_i)^2)(\sum_{i=1}^{n}(z_i - y_i)^2))^{\frac{1}{2}}$

$\geq \sum_{i=1}^{n}(y_i - x_i)^2 + \sum_{i=1}^{n}(z_i - y_i)^2 + 2\sum_{i=1}^{n}(y_i - x_i)(z_i - y_i)$

[using Cauchy's inequality $(\sum a_i^2)(\sum b_i^2) \geq (\sum a_i b_i)^2$]

$= \sum_{i=1}^{n}((y_i - x_i) + (z_i - y_i))^2$

$= \sum_{i=1}^{n}(z_i - x_i)^2$

$= (d(x, z))^2$.

So d is a metric on $\mathbf{R}^n$.

3. (a) $d(x,y) = |y_1 - x_1| + |y_2 - x_2| \geq 0$ for all x, y in $\mathbf{R}^2$.
$d(x,y) = 0 \iff |y_1 - x_1| + |y_2 - x_2| = 0 \iff |y_1 - x_1| = |y_2 - x_2| = 0$
$\iff y = x$.

(b) $d(y,x) = |x_1 - y_1| + |x_2 - y_2| = |y_1 - x_1| + |y_2 - x_2| = d(x,y)$
for all x, y in $\mathbf{R}^2$.

(c) For all x, y, z in $\mathbf{R}^2$ we have

$$
\begin{aligned}
d(x,y) + d(y,z) &= |y_1 - x_1| + |y_2 - x_2| + |z_1 - y_1| + |z_2 - y_2| \\
&\geq |(y_1 - x_1) + (z_1 - y_1)| + |(y_2 - x_2) + (z_2 - y_2)| \\
&= |z_1 - x_1| + |z_2 - x_2| \\
&= d(x,z).
\end{aligned}
$$

So d is a metric on $\mathbf{R}^2$.

4. (a) $d(x,y) = \max\{|y_1 - x_1|, |y_2 - x_2|\} \geq 0$ for all x, y in $\mathbf{R}^2$.
$d(x,y) = \max\{|y_1 - x_1|, |y_2 - x_2|\} = 0 \iff |y_1 - x_1| = |y_2 - x_2| = 0$
$\iff y = x$.

(b) Clearly $d(y,x) = d(x,y)$ for all x, y in $\mathbf{R}^2$.

(c) For all x, y, z in $\mathbf{R}^2$ we have

$$d(x,y) + d(y,z) = \max\{|y_1 - x_1|, |y_2 - x_2|\} + \max\{|z_1 - y_1|, |z_2 - y_2\}.$$

So we have

$$d(x,y) + d(y,z) \geq |y_1 - x_1| + |z_1 - y_1| \geq |z_1 - x_1|$$

and similarly

$$d(x,y) + d(y,z) \geq |z_2 - x_2|.$$

Thus $d(x,y) + d(y,z) \geq \max\{|z_1 - x_1|, |z_2 - x_2|\} = d(x,z)$.
So d is a metric on $\mathbf{R}^2$.

5. (a) $d(f,g)$ is the supremum of a set of non-negative real numbers.
So $d(f,g) \geq 0$ for all f, g in E.
$d(f,g) = 0 \iff \{|g(x) - f(x)|\}_{x \in A} = \{0\} \iff g(x) = f(x)$ for all x
in $A \iff g = f$.

(b) $d(g,f) = d(f,g)$ for all f, g in E.

(c) Let f, g, h be any three elements of E.

For each element x_1 of A we have

$$|h(x_1) - f(x_1)| \leq |h(x_1) - g(x_1)| + |g(x_1) - f(x_1)|$$
$$\leq \sup_{x \in A} |h(x) - g(x)| + \sup_{x \in A} |g(x) - f(x)|.$$

So

$$\sup_{x \in A} |h(x) - f(x)| \leq \sup_{x \in A} |h(x) - g(x)| + \sup_{x \in A} |g(x) - f(x)|,$$

i.e. $d(f, h) \leq d(f, g) + d(g, h)$.

Thus d is a metric on E.

6. (a) $d(x, y) \geq 0$ for all x, y in E and $d(x, y) = 0 \Longleftrightarrow x = y$.

(b) $d(y, x) = d(x, y)$ for all x, y in E.

(c) If $x = y$ or $y = z$ then $d(x, z) = d(x, y) + d(y, z)$.

Suppose, then, that $x \neq y$ and $y \neq z$; then we have $d(x, y) + d(y, z) = 2 > d(x, z)$.

Thus d is a metric on E.

7. (a) $d(f, g) \geq 0$ for all f, g in E.

If $f \neq g$ there is a point a of I such that $|g(a) - f(a)| \neq 0$. Since f and g are continuous there is a subinterval J of I containing a and a positive real number m such that $|g(x) - f(x)| \geq m$ for all points x of J. It follows that

$$\int_I |g - f| \geq m.l(J) > 0,$$

where $l(J)$ is the length of the interval J.

So $d(f, g) = 0 \Longleftrightarrow g = f$.

(b) $d(g, f) = d(f, g)$ for all f, g in E.

(c) For all f, g, h in E we have

$$\begin{aligned}
d(f, g) + d(g, h) &= \int_I |g - f| + \int_I |h - f| \\
&= \int_I (|g - f| + |h - g|) \\
&\geq \int_I |h - f| \\
&= d(f, h).
\end{aligned}$$

Thus d is a metric on E.

8. (a) $\emptyset \in \mathbf{P}(E)$ and $E \in \mathbf{P}(E)$.

(b) If $(X_i)_{i \in I}$ is a family of subsets of E then $\bigcup_{i \in I} X_i \in \mathbf{P}(E)$.

(c) If $(X_i)_{i \in I}$ is a finite family of subsets of E then $\bigcap_{i \in I} X_i \in \mathbf{P}(E)$.

9. Trivial!

10. (a) $\emptyset \in T_p$ and since $p \in E$ we have $E \in T_p$.

(b) Let $(X_i)_{i \in I}$ be a family of sets in T_p. If all the sets X_i are empty, then $\bigcup_{i \in I} X_i = \emptyset \in T_p$. If at least one of the sets X_i, say X_{i_0}, is non-empty, then $p \in X_{i_0}$ and so $p \in \bigcup_{i \in I} X_i$; so $\bigcup_{i \in I} X_i \in T_p$.

(c) Let $(X_i)_{i \in I}$ be a finite family of sets in T_p. If at least one of the sets X_i is empty, then $\bigcap_{i \in I} X_i = \emptyset \in T_p$. If all the sets X_i are non-empty, we have $p \in X_i$ for all i in I. So $p \in \bigcap_{i \in I} X_i$; hence $\bigcap_{i \in I} X_i \in T_p$.

11. (a) $E \in T_{-p}$ and since $p \notin \emptyset$ we have $\emptyset \in T_{-p}$.

(b) Let $(X_i)_{i \in I}$ be a family of sets in T_{-p}. If at least one of the sets $X_i = E$ then $\bigcup_{i \in I} X_i = E \in T_{-p}$. If all the sets X_i are proper subsets of E then $p \notin X_i$ for all i in I. So $p \notin \bigcup_{i \in I} X_i$ and hence $\bigcup_{i \in I} X_i \in T_{-p}$.

(c) Let $(X_i)_{i \in I}$ be a finite family of sets in T_{-p}. If all the sets $X_i = E$ then $\bigcap_{i \in I} X_i = E \in T_{-p}$. If at least one of the sets, say X_{i_0}, is a proper subset, then $p \notin X_{i_0}$ and so $p \notin \bigcap_{i \in I} X_i$ (which is a subset of X_{i_0}). Thus $\bigcap_{i \in I} X_i \in T_{-p}$.

12. (a) $\emptyset \in T$ and since $C_E(E) = \emptyset$ is finite, $E \in T$.

(b) Let $(X_i)_{i \in I}$ be a family of sets in T. If all the sets $X_i = \emptyset$ then $\bigcup_{i \in I} X_i = \emptyset \in T$; if at least one of the sets X_i, say X_{i_0}, is non-empty, we have $C_E(\bigcup_{i \in I} X_i) = \bigcap_{i \in I} C_E(X_i) \subseteq C_E(X_{i_0})$, which is finite. So $\bigcup_{i \in I} X_i \in T$.

(c) Let $(X_i)_{i \in I}$ be a finite family of sets in T. If at least one of the sets X_i is empty then $\bigcap_{i \in I} X_i = \emptyset \in T$. If all the sets X_i are non-empty, we have $C_E(\bigcap_{i \in I} X_i) = \bigcup_{i \in I} C_E(X_i)$ which is a finite union of finite sets, hence finite. So $\bigcap_{i \in I} X_i \in T$.

13. (a) Since $p \notin \emptyset$ we have $\emptyset \in T$; since $C_E(E) = \emptyset$ is finite, we have $E \in T$.

(b) Let $(X_i)_{i \in I}$ be a family of sets in T. If $p \notin X_i$ for all i in I then

$p \notin \bigcup_{i \in I} X_i$; so $\bigcup_{i \in I} X_i \in T$. If there is at least one of the sets X_i, say X_{i_0}, such that $p \in X_{i_0}$, then $C_E(X_{i_0})$ is finite and hence $C_E(\bigcup_{i \in I} X_i)$ $= \bigcap_{x \in I} C_E(X_i) \subseteq C_E(X_{i_0})$ is finite also; hence $\bigcup_{i \in I} X_i \in T$.

(c) Let $(X_i)_{i \in I}$ be a finite family of sets in T. If $p \in X_i$ for all i in I then $C_E(X_i)$ is finite for all i in I and hence $C_E(\bigcap_{i \in I} X_i) =$ $\bigcup_{i \in I} C_E(X_i)$ is finite; so $\bigcap_{i \in I} X_i \in T$. If there is at least one of the sets X_i, say X_{i_0}, such that $p \notin X_{i_0}$, then $p \notin \bigcap_{i \in I} X_i$ and so $\bigcap_{i \in I} X_i \in T$.

14. Let $E = \{a, b, c\}$. There are 29 topologies on E:

    1. $\{E, \emptyset\}$.
    2. $\mathbf{P}(E)$.
    3-5. Three like $\{E, \emptyset, \{a\}\}$.
    6-8. Three like $\{E, \emptyset, \{a, b\}\}$.
    9-14. Six like $\{E, \emptyset, \{a\}, \{a, b\}\}$.
    15-17. Three like $\{E, \emptyset, \{a\}, \{b, c\}\}$.
    18-20. Three like $\{E, \emptyset, \{a\}, \{b\}, \{a, b\}\}$.
    21-23. Three like $\{E, \emptyset, \{a\}, \{a, b\}, \{a, c\}\}$.
    24-29. Six like $\{E, \emptyset, \{a\}, \{b\}, \{a, b\}, \{a, c\}\}$.

15. (a) $\emptyset \in T_d$ and $E \in T_d$.

(b) Let $(X_i)_{i \in I}$ be a family of sets in T_d. Let a be any point of $\bigcup_{i \in I} X_i$. Then there is an index i_0 in I such that $a \in X_{i_0}$. Since $X_{i_0} \in T_d$ there is a positive real number r such that $V_d(a, r) \subseteq X_{i_0}$. Then $V_d(a, r) \subseteq \bigcup_{i \in I} X_i$. So $\bigcup_{i \in I} X_i \in T_d$.

(c) Let $(X_i)_{i \in I}$ be a finite family of sets in T_d. Let a be any point of $\bigcap_{i \in I} X_i$. Then for each index i in I there is a positive real number r_i such that $V_d(a, r_i) \subseteq X_i$. Let r be the least of the numbers r_i; r is a positive real number. Then $V_d(a, r) \subseteq \bigcap_{i \in I} X_i$. So $\bigcap_{i \in I} X_i \in T_d$

16. Let X be any subset of E; let a be any point of X. Then $V_d(a, 1) = \{a\} \subseteq X$. Thus $X \in T_d$. It follows that $T_d = \mathbf{P}(E)$, the discrete topology.

17. Let q be any point of $V_d(a, r)$.

Let $s = r - d(a, q)$. Then we have $V_d(q, s) \subseteq V_d(a, r)$, since, if x is any point of $V_d(q, s)$, we have $d(a, x) \leq d(a, q) + d(q, x) < d(a, q) + s = r$.

18. (1)(a) For all points x, y in E we have $d'(x,y) = \min\{1, d(x,y)\} \geq 0$.

$d'(x,y) = 0 \iff \min\{1, d(x,y)\} = 0 \iff d(x,y) = 0 \iff x = y$.

(b) For all x, y in E we have $d'(y,x) = d'(x,y)$.

(c) Let x, y, z be any elements of E.

Consider the sum $d'(x,y) + d'(y,z)$. If $d'(x,y) = 1$ or $d'(y,z) = 1$ then we have $d'(x,y) + d'(y,z) \geq 1 \geq d'(x,z)$. Otherwise we have $d'(x,y) = d(x,y)$ and $d'(y,z) = d(y,z)$, whence $d'(x,y) + d'(y,z) \geq d(x,z) \geq d'(x,z)$.

So d' is a metric on E.

(2) Let $U \in T_d$; let p be any point of U. Then there is a positive real number r such that $V_d(p,r) \subseteq U$. Let $r' = \min\{r,1\}$. Then $V_{d'}(p,r') = V_d(p,r') \subseteq V_d(p,r) \subseteq U$. Thus $U \in T_{d'}$.

Conversely, suppose $U' \in T_{d'}$; let p be any point of U'. Then there is a positive real number r' such that $V_{d'}(p,r') \subseteq U'$. Clearly we may suppose that $r' \leq 1$. Then $V_d(p,r') = V_{d'}(p,r') \subseteq U'$. So $U' \in T_d$.

19. Let d_2, d_3, d_4 be the metrics on $\mathbf{R}^2$ introduced in Exercises 2, 3, 4.

Let X be any subset of $\mathbf{R}^2$.

(1) Suppose X is d_2-open, i.e. open relative to the metric d_2.

Let p be any point of X. Then there is a positive real number r such that $V_{d_2}(p,r) \subseteq X$. Then we have $V_{d_3}(p,r) \subseteq V_{d_2}(p,r)$ and $V_{d_4}(p,r/\sqrt{2}) \subseteq V_{d_2}(p,r)$.

So $V_{d_3}(p,r) \subseteq X$ and $V_{d_4}(p,r/\sqrt{2}) \subseteq X$. Thus X is both d_3- and d_4-open.

(2) Suppose X is d_3-open.

Let p be any point of X. Then there is a positive real number r such that $V_{d_3}(p,r) \subseteq X$. Since $V_{d_2}(p,r/\sqrt{2}) \subseteq V_{d_3}(p,r)$ we have $V_{d_2}(p,r/\sqrt{2}) \subseteq X$. Thus X is d_2-open, and hence also d_4-open.

(3) Suppose X is d_4-open.

Let p be any point of X. Then there is a positive real number r such that $V_{d_4}(p,r) \subseteq X$. Since $V_{d_2}(p,r) \subseteq V_{d_4}(p,r)$ we have $V_{d_2}(p,r) \subseteq X$. Thus X is d_2-open and hence also d_3-open.

Hence $T_{d_2} = T_{d_3} = T_{d_4}$.

20. If p is a pseudometric on E then, for each point a of E and each positive real number r, we define the r-ball with centre a to be the set $V_p(a, r) = \{x \in E : p(a, x) < r\}$. Then we say that a subset U of E is open relative to the pseudometric p if, for every point t of U, there is a positive real number r_t such that $V_p(t, r_t) \subseteq U$. The set T_p of all subsets which are open relative to p is shown to be a topology as in Exercise 15 and is called the topology induced by p.

21. Just like Exercise 20.

22. (a) $p(x, y) = |f(x) - f(y)| \geq 0$ and $p(x, x) = 0$.

(b) $p(x, y) = |f(x) - f(y)| = |f(y) - f(x)| = p(y, x)$.

(c) $p(x, y) + p(y, z) = |f(x) - f(y)| + |f(y) - f(z)| \leq |f(x) - f(z)| = p(x, z)$.

23. (a) If $x \leq y$ then $q(x, y) = y - x \geq 0$; if $y < x$ then $q(x, y) = 1 > 0$.

If $x = y$ then $q(x, y) = 0$; if $q(x, y) = 0$ we cannot have $y < x$, so $x \leq y$ and so $0 = q(x, y) = y - x$; thus $x = y$.

(b) Let x, y, z be points of E. If any two of these points coincide we certainly have $q(x, z) \leq q(x, y) + q(y, z)$.

Suppose now that all three points are distinct. There are six cases to consider: $x < y < z$; $x < z < y$; $y < x < z$; $y < z < x$; $z < x < y$; $z < y < x$. In each case we have $q(x, z) \leq q(x, y) + q(y, z)$.

24. Let X and Y be R-classes; let x_1, x_2 be elements of E in X, y_1, y_2 elements of E in Y. Then we have $p(x_1, x_2) = p(x_2, x_1) = 0$ and $p(y_1, y_2) = p(y_2, y_1) = 0$. Hence $p(x_1, y_1) \leq p(x_1, x_2) + p(x_2, y_1) = p(x_2, y_1) \leq p(x_2, y_2) + p(y_2, y_1) = p(x_2, y_2)$ and $p(x_2, y_2) \leq p(x_2, y_1) + p(y_1, y_2) = p(x_2, y_1) \leq p(x_2, x_1) + p(x_1, y_1) = p(x_1, y_1)$ So $p(x_1, y_1) = p(x_2, y_2)$.

Thus the definition of p^* suggested, i.e. $p^*(X, Y) = p(x, y)$ where $x \in X$ and $y \in Y$, is independent of the choice of x and y. Clearly $p^*(\eta(x), \eta(y)) = p(x, y)$.

To show that p^* is a metric, we proceed as follows (using the obvious convention that x, y, z are elements of E chosen from the R-classes X, Y, Z respectively).

(a) $p^*(X, Y) = p(x, y) \geq 0$; $p^*(X, X) = p(x, x) = 0$.

If $p^*(X, Y) = 0$ then $p(x, y) = 0$, so $(x, y) \in R$ and hence $X = Y$.

(b) $p^*(X, Y) = p(x, y) = p(y, x) = p^*(Y, X)$.

(c) $p^*(X, Y) + p^*(Y, Z) = p(x, y) + p(y, z) \geq p(x, z) = p^*(X, Z)$.

25. Let x, y, z be three points of E. Let $a = u(x, y)$, $b = u(y, z)$, $c = u(x, z)$. Suppose $c = \max\{a, b, c\}$. Now $c \leq \max\{a, b\}$ by **D3u** and $\max\{a, b\} \leq \max\{a, b, c\} = c$. So $c = \max\{a, b\} = a$ say; thus we have $c = a$ and $b \leq c$.

26. (a) For all x, y we have $u(x, y) \geq 0$ and $u(x, x) = 0$.

If $x \neq y$ then $u(x, y) > 0$; so if $u(x, y) = 0$ we have $x = y$.

(b) If $y - x = p^\alpha m/n$ then $x - y = p^\alpha(-m)/n$; so we have $u(y, x) = p^{-\alpha} = u(x, y)$.

(c) Suppose $y - x = p^\alpha m/n$ and $z - y = p^\beta q/r$. Then $z - x$ has the form $p^\gamma s/t$ where $\gamma \geq \min\{\alpha, \beta\}$. So $u(x, z) = p^{-\gamma} \leq \max\{p^{-\alpha}, p^{-\beta}\} = \max\{u(x, y), u(y, z)\}$.

27. (1) Suppose B is a base for T.

Let U be a set in T, x any point of U. Then there is a family $(W_i)_{i \in I}$ of sets in B such that $\bigcup_{i \in I} W_i = U$. Since $x \in U$ there is an index i_0 in I such that $x \in W_{i_0}$ and of course $W_{i_0} \subseteq U$.

(2) Conversely, suppose the condition holds.

Let U be any set in T. For every point x of U there exists a set W_x in B such that $x \in W_x$ and $W_x \subseteq U$. Then $U = \bigcup_{x \in U} W_x$. So B is a base for T.

28. Let T be the set of all subsets of E which are unions of families of sets in B.

If we can show that T is a topology it is clear that B is a base for T and that it is the only topology of which B is base.

T certainly enjoys properties **T1** and **T2** for a topology.

To show that T satisfies condition **T3**, it is clearly sufficient to show that, if U_1 and U_2 are sets in T, then $U_1 \cap U_2$ is also in T.

So let x be any point in $U_1 \cap U_2$. Since $x \in U_1$ there is a set W_1 in B such that $x \in W_1 \subseteq U_1$; since $x \in U_2$ there is a set W_2 in B such that $x \in W_2 \subseteq U_2$. By hypothesis there is a set W_x in B such that $x \in W_x$ and $W_x \subseteq W_1 \cap W_2 \subseteq U_1 \cap U_2$. It follows that $U_1 \cap U_2 = \bigcup_{x \in U_1 \cap U_2} W_x$; so $U_1 \cap U_2 \in T$.

29. The conditions of Exercise 28 are clearly satisfied by X.

30. Let a, b, c, d be elements of E such that $a \leq b$ and $c \leq d$. If $\max\{a, c\} \geq \min\{b, d\}$ then $\{t : a < t < b\}$ and $\{t : c < t < d\}$ are disjoint; otherwise these open intervals intersect in the open interval $\{t : \max\{a, c\} < t < \min\{b, d\}\}$. So the conditions of Exercise 28 are again satisfied.

31. $\{t : t > a\} \cap \{t : t > b\} = \{t : t > \max\{a, b\}\}$.
Again the conditions of Exercise 28 are satisfied.

32. Let S' be the collection of subsets which are unions of families of intersections of finite families of sets in S, together with E.
  Then S' is a topology on E which includes S. Hence $S' \supseteq T(S)$. On the other hand, since $T(S)$ is a topology including S, it must contain all the sets in S'. So $T(S) \supseteq S'$.
  Hence $T(S) = S'$ as required.

33. Let $T = \{U \in \mathbf{P}(E) : C_E(U) \in K\}$.
  Then T is a topology on E and F is T-closed $\Longleftrightarrow C_E(F) \in T$ $\Longleftrightarrow C_E(C_E(F)) = F \in K$.

34. (1) Let $G = \bigcap_{i \in I} U_i$ (where I is countable and all the sets U_i are T-open) be a G_δ-set. Then $C_E(G) = C_E(\bigcap_{i \in I} U_i) = \bigcup_{i \in I} C_E(U_i)$, which is an F_σ-set, since all the sets $C_E(U_i)$ are T-closed.
  Similarly the complement of an F_σ-set is a G_δ-set.
  (2) For each natural number n let $F_n = \bigcup_{p \leq n} K_p$. Each of the sets F_n is T-closed. Further, $F_n \subseteq F_{n+1}$ for all natural numbers n and $\bigcup_{n \in \mathbf{N}} F_n = \bigcup_{n \in \mathbf{N}} K_n = K$.

35. (1) $\emptyset \in T^*$ by hypothesis; $E \in T$, so $E^* = E \cup \{p\} \in T^*$.
  (2) Let $(U_i^*)_{i \in I}$ be a family of sets in T^*.
  If all the sets U_i^* are empty then $\bigcup_{i \in I} U_i^* = \emptyset \in T^*$. Otherwise, let $I' = \{i \in I : U_i^* \neq \emptyset\}$; for each index i in I' there is a set U_i in T such that $U_i^* = U_i \cup \{p\}$. Then $\bigcup_{i \in I} U_i^* = \bigcup_{i \in I'} U_i^* = (\bigcup_{i \in I'} U_i) \cup \{p\} \in T^*$.
  (3) Let $(U_j^*)_{j \in J}$ be a finite family of sets in T^*.
  If at least one of the sets U_j^* is empty, then $\bigcap_{j \in J} U_j^* = \emptyset \in T^*$.

Otherwise, for each index j, in J there is a set U_j in T such that $U_j^* = U_j \cup \{p\}$. Then $\bigcap_{j \in J} U_j^* = (\bigcap_{j \in J} U_j) \cup \{p\} \in T^*$.

Thus T^* is a topology on E^*.

Let X be a subset of E. Then X is T^*-closed $\iff C_{E^*}(X) \in T^*$ $\iff (C_E(X)) \cup \{p\} \in T^* \iff X$ is T-closed.

Finally we have $U \in T_p$
$\iff U = \emptyset$ or $U = X \cup \{p\}$ where X is any subset of $C_E\{p\}$
$\iff U = \emptyset$ or $U = $ (a set of the discrete topology on $C_E\{p\}) \cup \{p\}$
$\iff U \in$ the closed extension topology of the discrete topology on $C_E\{p\}$.

36. (1) $0 \notin \emptyset$, so $\emptyset \in T$; $(-1, 1) \subset E$, so $E \in T$.

(2) Let $(U_i)_{i \in I}$ be a family of sets in T.

Suppose $0 \notin U_i$ for all indices i in I. Then $0 \notin \bigcup_{i \in I} U_i$ and hence $\bigcup_{i \in I} U_i \in T$. Otherwise there is an index i_0 in I such that $(-1, 1) \subseteq U_{i_0}$. Then $(-1, 1) \subseteq \bigcup_{i \in I} U_i$ and hence $\bigcup_{i \in I} U_i \in T$.

(3) Let $(U_j)_{j \in J}$ be a finite family of sets in T.

Suppose $(-1, 1) \subseteq U_j$ for all indices j in J. Then $(-1, 1) \subseteq \bigcap_{j \in J} U_j$ and hence $\bigcap_{j \in J} U_j \in T$. Otherwise there is an index j_0 in J such that $0 \notin U_{j_0}$. Then $0 \notin \bigcap_{j \in J} U_j$ and hence $\bigcap_{j \in J} U_j \in T$.

Thus T is a topology on E.

Let X be any subset of E.

Then X is T-closed
$$\iff C_E(X) \in T$$
$$\iff (-1, 1) \subseteq C_E(X) \text{ or } 0 \notin C_E(X)$$
$$\iff X \subseteq C_E((-1, 1)) \text{ or } 0 \in X$$
$$\iff X = \emptyset \text{ or } \{1\} \text{ or } \{-1\} \text{ or } \{-1, 1\} \text{ or else } 0 \in X.$$

37. Let V be any T_d-neighbourhood of p. Then there is a T_d-open set U such that $p \in U \subseteq V$.

Since U is T_d-open and $p \in U$ there is a positive real number r such that $V_d(p, r) \subseteq U$. There exists a positive rational number q such that $q < r$. Then $V_d(p, q) \subseteq V_d(p, r) \subseteq U \subseteq V$. Thus the set of balls with centre p and positive rational radii forms a neighbourhood base.

38. (1) Let U be a T-open set, p a point of U. Since U is a T-open set containing p and included in U, it follows that U is a T-neighbourhood of p.

(2) Conversely, suppose U is a T-neighbourhood of each of its points. Then for each point x of U there is a T-open set U_x such that $x \in U_x$ and $U_x \subseteq U$. Then $U = \bigcup_{x \in U} U_x$ and hence is T-open.

39. The finest topology on E is the discrete topology $\mathbf{P}(E)$; the coarsest topology on E is the trivial topology $\{E, \emptyset\}$.

Suppose T is finer than T'. Let x be a point of E, V' a T'-neighbourhood of x. Then there is a T'-open set U' such that $x \in U'$ and $U' \subseteq V'$. Since $T' \subseteq T$, U' is T-open. Hence V' is a T-neighbourhood of x.

Conversely, suppose that for every point x of E every T'-neighbourhood is a T-neighbourhood. Let U' be a T'-open set. Then U' is a T'-neighbourhood of each of its points and hence a T-neighbourhood of each of its points. Thus U' is T-open. So $T' \subseteq T$, i.e. T is finer than T'.

40. Let $T = \{U \in \mathbf{P}(E) : U \in N(x) \text{ for all points } x \text{ of } U\}$.

(1) E and $\emptyset$ belong to T.

(2) Let $(U_i)_{i \in I}$ be a family of sets in T; let $x \in \bigcup_{i \in I} U_i$. Then there is an index i_0 in I such that $x \in U_{i_0}$. Since U_{i_0} is in T we have $U_{i_0} \in N(x)$. Since $\bigcup_{i \in I} U_i \supseteq U_{i_0}$ it follows that $\bigcup_{i \in I} U_i \in N(x)$. Hence $\bigcup_{i \in I} U_i \in T$.

(3) Let $(U_j)_{j \in J}$ be a finite family of sets in T; let $x \in \bigcap_{j \in J} U_j$. Then for each index j in J we have $U_j \in N(x)$. It follows that for each $x \in \bigcap_{j \in J} U_j$ we have $\bigcap_{j \in J} U_j \in N(x)$. Hence we have $\bigcap_{j \in J} U_j \in T$.

So T is a topology on E.

Let x be any point of E, V any T-neighbourhood of x. Then there is a set U in T such that $x \in U$ and $U \subseteq V$. Since $U \in N(x)$ (because $U \in T$), it follows that $V \in N(x)$.

Conversely, let x be any point of E, V any set in $N(x)$. Let $U = \{y \in E : V \in N(y)\}$. Clearly $x \in U$. Next, $U \subseteq V$; for if $y \in U$ we have $V \in N(y)$ and hence $y \in V$. Finally, $U \in T$. To see this, let t be any point of U. Since $V \in N(t)$ there is a set W in $N(t)$ such that $V \in N(w)$ for all points w of W. Since $V \in N(w)$ for all points w of W, it follows that $W \subseteq U$. Since $W \in N(t)$ it follows that $U \in N(t)$. Thus $U \in T$ and so V is a T-neighbourhood of x.

The uniqueness of T is clear.

41. Int $A \subseteq A$ and Int $B \subseteq B \implies$ Int $A \cap$ Int $B \subseteq A \cap B$. Since Int $A \cap$ Int B is an open set included in $A \cap B$ it follows that Int $A \cap$ Int $B \subseteq$ Int $(A \cap B)$.

Conversely Int $(A \cap B)$ is an open subset of $A \cap B$, hence of A. So Int $(A \cap B) \subseteq$ Int A. Similarly Int $(A \cap B) \subseteq$ Int B. So we have Int $(A \cap B) \subseteq$ Int $A \cap$ Int B. Hence Int $(A \cap B) =$ Int $A \cap$ Int B.

Similarly Cl $(A \cup B) =$ Cl $A \cup$ Cl B.

If $A \subseteq B$, Cl B is a closed set which includes B, hence includes A. So Cl $B \supseteq$ Cl A.

Similarly, if $A \subseteq B$ we have Int $A \subseteq$ Int B.

42. Since Int $A \subseteq A$ we have $C_E(\text{Int } A) \supseteq C_E(A)$. Since Int A is open, $C_E(\text{Int } A)$ is closed and so $C_E(\text{Int } A) \supseteq$ Cl $(C_E(A))$.

Conversely, Cl $(C_E(A))$ is a closed set which includes $C_E(A)$. So $C_E(\text{Cl }(C_E(A)))$ is an open set included in A. It follows that we have $C_E(\text{Cl }(C_E(A))) \subseteq$ Int A and so $C_E(\text{Int } A) \subseteq$ Cl $(C_E(A))$. Hence we have Cl $(C_E(A)) = C_E(\text{Int } A)$.

43. Let (E, T) be **R** with the topology induced by the usual metric.

For each natural number $n \geq 1$ let $A_n = \{x \in \mathbf{R} : -1/n < x\}$; then Int $A_n = A_n$ for all such n. Then we have $\bigcap_{n \geq 1}$ Int $A_n = \bigcap_{n \geq 1} A_n = \{x \in \mathbf{R} : x \geq 0\} \supset$ Int $(\bigcap_{n \geq 1} A_n) = \{x \in \mathbf{R} : x > 0\}$.

44. It follows from condition (3) that if X and Y are subsets of E such that $X \subseteq Y$ then $\kappa(X) \subseteq \kappa(Y)$.

(1) $\kappa(C_E(\emptyset)) = \kappa(E) = E = C_E(\emptyset)$. So $\emptyset \in T_\kappa$.

$\kappa(C_E(E)) = \kappa(\emptyset) = \emptyset = C_E(E)$. So $E \in T_\kappa$.

(2) Let $(U_i)_{i \in I}$ be a family of sets in T_κ. Then $\kappa(C_E(\bigcup_{i \in I} U_i)) \supseteq C_E(\bigcup_{i \in I} U_i)$. On the other hand, we have, for all j in I,

$$\kappa(C_E(\bigcup_{i \in I} U_i)) = \kappa(\bigcap_{i \in I}(C_E(U_i))) \subseteq \kappa(C_E(U_j)) = C_E(U_j).$$

Hence we have $\kappa(C_E(\bigcup_{i \in I} U_i)) \subseteq \bigcap_{i \in I}(C_E(U_i)) = C_E(\bigcup_{i \in I} U_i)$. Thus $\kappa(C_E(\bigcup_{i \in I} U_i)) = C_E(\bigcup_{i \in I} U_i)$ and so $\bigcup_{i \in I} U_i \in T_\kappa$.

(3) Let $(U_j)_{j \in J}$ be a finite family of sets in T_κ.

Then $\kappa(C_E(\bigcap_{j \in J} U_j)) = \kappa(\bigcup_{j \in J}(C_E(U_j))) = \bigcup_{j \in J} \kappa(C_E(U_j))$ (using condition (3)) $= \bigcup_{j \in J}(C_E(U_j)) = C_E(\bigcap_{j \in J} U_j)$. So we have $\bigcap_{j \in J} U_j \in T_\kappa$.

Thus T_κ is a topology on E.

Let F be any subset of E. Then F is T_κ-closed $\iff C_E(F) \in T_\kappa$ $\iff \kappa(C_E(C_E(F))) = C_E(C_E(F)) \iff \kappa(F) = F$.

Now let X be any subset of E. Since $\kappa(X) \supseteq X$ and $\kappa(\kappa(X)) = \kappa(X)$, we see that $\kappa(X)$ is a T_κ-closed subset which includes X. So $\kappa(X) \supseteq \text{Cl}_{T_\kappa}(X)$. On the other hand, since $\text{Cl}_{T_\kappa}(X)$ is T_κ-closed and $\text{Cl}_{T_\kappa}(X) \supseteq X$ we deduce that $\text{Cl}_{T_\kappa}(X) = \kappa(\text{Cl}_{T_\kappa}(X)) \supseteq \kappa(X)$. Thus $\text{Cl}_{T_\kappa}(X) = \kappa(X)$.

45. The verification that κ satisfies the conditions of Exercise 44 is straightforward.

$U \in T_\kappa \iff \kappa(C_E(U)) = C_E(U) \iff C_E(U)$ is either finite or $= E$ $\iff U \in$ the finite complement topology.

46. Int $A = \text{Cl}\,A = A$, Fr $A = \emptyset$ for all subsets A of E.
A is nowhere dense $\iff \text{Int Cl}\,A = \emptyset \iff A = \emptyset$.

So if E is non-empty it cannot be expressed as a union of nowhere dense subsets. Thus E is second category.

47. We have
$$\text{Int } A = \begin{cases} \emptyset & \text{if } A \neq E \\ E & \text{if } A = E. \end{cases}$$
$$\text{Cl } A = \begin{cases} \emptyset & \text{if } A = \emptyset \\ E & \text{if } A \neq \emptyset. \end{cases}$$
$$\text{Fr } A = \begin{cases} \emptyset & \text{if } A = E \text{ or } \emptyset \\ E & \text{otherwise.} \end{cases}$$
A is nowhere dense $\iff \text{Int Cl}\,A = \emptyset \iff \text{Cl}\,A \neq E \iff A = \emptyset$.
As in Exercise 46, E is second category.

48. $\text{Cl}\,P_r = \{x \in \mathbf{R} : x \leq r\}$; $\text{Int Cl}\,P_r = \emptyset$. So P_r is nowhere dense. $\mathbf{R} = \bigcup_{n \in \mathbf{N}} P_n$. So $\mathbf{R}$ is meagre.

49. (1) Let x be an interior point of A.

Then there is a T-open set U such that $x \in U$ and $U \subseteq A$. Hence $U \subseteq \text{Int } A$ and so $x \in \text{Int } A$.

(2) Conversely, if $x \in \text{Int } A$, then of course Int A is T-open and Int $A \subseteq A$. From this it follows at once that x is an interior point of A.

50. Let $x \in \text{Cl}\,A$.

If x is not an adherent point of A, there is a neighbourhood V of x such that $V \cap A = \emptyset$. V includes an open set U which contains x, and of course $U \cap A = \emptyset$, so that $C_E(U) \supseteq A$. Since $C_E(U)$ is a closed set which includes A, we have $C_E(U) \supseteq \operatorname{Cl} A$ and so $x \notin U$, which is a contradiction. Thus $x \in \operatorname{Cl} A \Longrightarrow x$ is an adherent point of A.

Conversely, suppose $x \notin \operatorname{Cl} A$. Then there is a closed subset F such that $F \supseteq A$ and $x \notin F$. Then $C_E(F)$ is an open set, containing x, hence a neighbourhood of x, and $C_E(F) \cap A = \emptyset$. So x is not an adherent point of A. So x is an adherent point of $A \Longrightarrow x \in \operatorname{Cl} A$.

51. We have $X \subseteq \operatorname{Cl} X$ and $\operatorname{Fr} X = \operatorname{Cl} X \cap \operatorname{Cl} C_E(X) \subseteq \operatorname{Cl} X$. So $X \cup \operatorname{Fr} X \subseteq \operatorname{Cl} X$.

Conversely, let $x \in \operatorname{Cl} X$. If $x \notin \operatorname{Fr} X$ then there is a neighbourhood V of x which does not meet $C_E(X)$ and hence is included in X. Hence $x \in V \subseteq X$; so $x \in X$. Thus $\operatorname{Cl} X \subseteq X \cup \operatorname{Fr} X$. Hence $\operatorname{Cl} X = X \cup \operatorname{Fr} X$.

Now consider

$$\begin{aligned}
C_X(\operatorname{Fr} X) &= X \cap C_E \left(\operatorname{Cl} X \cap \operatorname{Cl} C_E X \right) \\
&= (X \cap C_E \operatorname{Cl} X) \cup (X \cap C_E \operatorname{Cl} C_E X \\
&= (X \cap \operatorname{Int} X) \cup (X \cap \operatorname{Int} C_E X) \\
&= X \cap \operatorname{Int} X = \operatorname{Int} X.
\end{aligned}$$

52. (1) $X \subseteq Y \Longrightarrow \operatorname{Cl} X \subseteq \operatorname{Cl} Y \Longrightarrow \operatorname{Int} \operatorname{Cl} X \subseteq \operatorname{Int} \operatorname{Cl} Y$, and so $\alpha(X) \subseteq \alpha(Y)$.

(2) If X is open, then since $X \subseteq \operatorname{Cl} X$ we have $X \subseteq \operatorname{Int} \operatorname{Cl} X = \alpha(X)$.

(3) Since $\alpha(X)$ is open we have $\alpha(X) \subseteq \alpha(\alpha(X))$.

Since $\operatorname{Int} \operatorname{Cl} X \subseteq \operatorname{Cl} X$ we have $\operatorname{Cl} \operatorname{Int} \operatorname{Cl} X \subseteq \operatorname{Cl} \operatorname{Cl} X = \operatorname{Cl} X$ and so $\operatorname{Int} \operatorname{Cl} \operatorname{Int} \operatorname{Cl} X \subseteq \operatorname{Int} \operatorname{Cl} X$, i.e. $\alpha(\alpha(X)) \subseteq \alpha(X)$.

So $\alpha(\alpha(X)) = \alpha(X)$.

(4) Suppose X and Y are disjoint open sets.

Suppose $\alpha(X) \cap \alpha(Y) \neq \emptyset$. Let t be a point of $\alpha(X) \cap \alpha(Y)$. Then $t \in \operatorname{Int} \operatorname{Cl} X \subseteq \operatorname{Cl} X$. So every open set containing t meets X. Thus $\alpha(Y) \cap X \neq \emptyset$. Let w be any element of $\alpha(Y) \cap X$. Then we have $w \in \operatorname{Int} \operatorname{Cl} Y \subseteq \operatorname{Cl} Y$. So every open set containing w meets Y. In particular X meets Y, which is a contradiction.

53. Let $(U_i)_{i \in I}$ be a finite family of regular open sets.

By Exercise 52(2) we have $\alpha(\bigcap_{i \in I} U_i) \supseteq \bigcap_{i \in I} U_i$ since $\bigcap_{i \in I} U_i$ is open. For each index i in I we have $U_i \supseteq \bigcap_{i \in I} U_i$ and so $\alpha(U_i) \supseteq \alpha(\bigcap_{i \in I} U_i)$ by Exercise 52(1). Since $\alpha(U_i) = U_i$ it follows that we have $\bigcap_{i \in I} U_i \supseteq \alpha(\bigcap_{i \in I} U_i)$. Thus $\bigcap_{i \in I} U_i$ is regular open.

Let $X = (0, \frac{1}{2})$, $Y = (\frac{1}{2}, 1)$. Then we have $\mathrm{Cl}\,(X) = [0, \frac{1}{2}]$ and so $\alpha(X) = (0, \frac{1}{2})$. Similarly $\alpha(Y) = (\frac{1}{2}, 1)$. $\mathrm{Cl}\,(X \cup Y) = [0, 1]$; then $\alpha(X \cup Y) = (0, 1) \supset \alpha(X) \cup \alpha(Y) = X \cup Y$. So $X \cup Y$ is not regular open.

54. Suppose $A \cup B = E$.

Let x be any point of E not in $\mathrm{Cl}\,A$. Then $x \notin A$ and hence $x \in B$. Further, there is a neighbourhood V of x which does not meet A and hence is included in B. Thus B is a neighbourhood of x, i.e. $x \in \mathrm{Int}\,B$. So $\mathrm{Cl}\,A \cup \mathrm{Int}\,B = E$.

55. (1) First

$$\begin{aligned} \mathrm{Fr}\,\mathrm{Cl}\,A &= \mathrm{Cl}\,\mathrm{Cl}\,A \cap \mathrm{Cl}\,C_E\,\mathrm{Cl}\,A \\ &= \mathrm{Cl}\,A \cap \mathrm{Cl}\,C_E\,\mathrm{Cl}\,A \\ &\subseteq \mathrm{Cl}\,A \cap \mathrm{Cl}\,C_E\,A = \mathrm{Fr}\,A \end{aligned}$$

Next we have

$$\begin{aligned} \mathrm{Fr}\,\mathrm{Int}\,A &= \mathrm{Cl}\,\mathrm{Int}\,A \cap \mathrm{Cl}\,C_E\,\mathrm{Int}\,A \\ &= \mathrm{Cl}\,\mathrm{Int}\,A \cap C_E\,\mathrm{Int}\,A \\ &= \mathrm{Cl}\,\mathrm{Int}\,A \cap \mathrm{Cl}\,C_E\,A \\ &\subseteq \mathrm{Cl}\,A \cap \mathrm{Cl}\,C_E\,A = \mathrm{Fr}\,A \end{aligned}$$

Take $A = (0, 1) \cup (1, 2) \cup \{3\}$. Then we have

$$\begin{aligned} \mathrm{Int}\,A &= (0, 1) \cup (1, 2), \\ \mathrm{Cl}\,A &= [0, 2] \cup \{3\}, \\ \mathrm{Fr}\,A &= \{0, 1, 2, 3\}, \\ \mathrm{Fr}\,\mathrm{Int}\,A &= \{0, 1, 2\}, \\ \mathrm{Fr}\,\mathrm{Cl}\,A &= \{0, 2, 3\}. \end{aligned}$$

(2) $\mathrm{Fr}\,(A \cup B) = \mathrm{Cl}\,(A \cup B) \cap \mathrm{Cl}\,(C_E\,(A \cup B))$

$$
\begin{aligned}
&= (\mathrm{Cl}\,A \cup \mathrm{Cl}\,B) \cap \mathrm{Cl}\,(C_E A \cap C_E B) \\
&= \{\mathrm{Cl}\,A \cap \mathrm{Cl}\,(C_E A \cap C_E B)\} \cup \{\mathrm{Cl}\,B \cap \mathrm{Cl}\,(C_E A \cap C_E B)\} \\
&\subseteq \{\mathrm{Cl}\,A \cap \mathrm{Cl}\,C_E A\} \cup \{\mathrm{Cl}\,B \cap \mathrm{Cl}\,C_E B\} \\
&= \mathrm{Fr}\,A \cup \mathrm{Fr}\,B.
\end{aligned}
$$

Take $A = (0, 1]$, $B = (1, 2)$. Then $A \cup B = (0, 2)$.

In this case we see that $\mathrm{Fr}\,A = \{0, 1\}$, $\mathrm{Fr}\,B = \{1, 2\}$ but we have $\mathrm{Fr}\,(A \cup B) = \{0, 2\}$.

56. Let x be any point of U.

Let V be any T-neighbourhood of x. Then $U \cap V$ is a T-neighbourhood of x. Since D is dense, x belongs to the closure of D and so we have $(U \cap V) \cap D \neq \emptyset$. Thus $V \cap (U \cap D) \neq \emptyset$ and hence $x \in \mathrm{Cl}\,(U \cap D)$. Thus $U \subseteq \mathrm{Cl}\,(U \cap D)$.

57. Suppose $\mathrm{Int}\,A \neq \emptyset$. Let D be a dense subset of E. If x is any point of $\mathrm{Int}\,A$ then $\mathrm{Int}\,A$ is a neighbourhood of x; since $x \in \mathrm{Cl}\,D = E$, $\mathrm{Int}\,A$ meets D and hence $D \cap A \neq \emptyset$.

Conversely, suppose $\mathrm{Int}\,A = \emptyset$. Then $\mathrm{Cl}\,(C_E(A)) = C_E(\mathrm{Int}\,A) = E$; so $C_E(A)$ is dense, but doesn't meet A.

58. Let F be a closed subset of E containing p. Then $C_E(F)$ is an open subset of E which does not contain p. Hence $C_E(F) = \emptyset$ and so $F = E$. Hence $\mathrm{Cl}\,\{p\} = E$.

Let F' be a proper closed subset of E. Then $C_E(F')$ is a non-empty open subset of E and hence contains p. Thus no open subset included in F' can contain p. So the only open set included in F' is $\emptyset$. Thus $\mathrm{Int}\,F' = \emptyset$.

Let X be any subset of E which contains p.

Let t be a point distinct from p, V any neighbourhood of t. Then V includes an open set U containing t. Since U is non-empty it contains p. Then $V' \cap X$ contains p. So t is a cluster point of X. $V = \{p, t\}$ is a neighbourhood of t such that $V \cap X$ is finite; so t is not an ω-accumulation point.

59. Let (E, T) be a second countable space. Let $(B_k)_{k \in K}$ be a countable base for T. For each index k in K let x_k be a point of B_k. Let $X = \{x_k\}_{k \in K}$; X is a countable subset of E. We claim that X is dense. To prove this, let p be any point of E, V any neighbourhood of p. Then V includes an open set, and hence a basic open set containing p—say $p \in B_{k_0} \subseteq V$. Then $V \cap X$ contains x_{k_0} and so is non-empty. So $p \in \operatorname{Cl} X$. Thus X is dense.

60. (1) As in Exercise 58 we have $\operatorname{Cl}\{p\} = E$; so $\{p\}$ is a countable dense subset of E. Thus (E, T_p) is separable.

For each point x of E the set $\{x, p\}$ is open and hence must be a union of sets from any base for T, and so, in fact, must belong to any base for T. There are uncountably many such sets. So there is no countable base for T.

(2) Let K be any countably infinite subset of E.

Then $C_E(\operatorname{Cl} K)$ is open and hence must either be empty or else have finite complement. But $C_E(C_E(\operatorname{Cl} K)) = \operatorname{Cl} K \supseteq K$ is infinite. Hence $C_E(\operatorname{Cl} K) = \emptyset$, i.e. $E = \operatorname{Cl} K$, so K is dense. Thus (E, T) is separable.

Suppose there were a countable base $(B_i)_{i \in I}$ for T. Let p be a point of E and let $J = \{i \in I : p \in B_i\}$. Then $\bigcap_{i \in J} B_i = \{p\}$ (for if $q \neq p$, $C_E\{q\}$ is a T-open set which contains p and so there is a set B_{i_0} with $i_0 \in J$ such that $p \in B_{i_0} \subseteq C_E\{q\}$; then $q \notin B_{i_0}$). It follows that $C_E\{p\}$ (which is uncountable) $= \bigcup_{i \in J} (C_E(B_i))$ (which is a countable union of finite sets, hence countable), and this is a contradiction. So (E, T) is not second countable.

61. Let $\{x_k\}_{k \in K}$ be a countable dense subset of E. Let $B = \{V_d(x_k, \frac{1}{n})\}_{k \in K, n \in \mathbf{P}}$, where $\mathbf{P}$ is the set of positive integers; B is a countable collection of open sets. We claim it is a base for the metric topology on E.

So let U be any T-open set. If x is any point of U there is a positive integer m such that $V_d(x, \frac{1}{m}) \subseteq U$. Since $\{x_k\}$ is dense, $V_d(x, \frac{1}{2m})$ contains one of the points x_k, say x_{k_0}. Then $x \in V_d(x_{k_0}, \frac{1}{2m}) \subseteq V_d(x, \frac{1}{m}) \subseteq U$.

62. Let $(B_k)_{k \in K}$ be a countable base for (E, T). Let p be any point of E. Let $J = \{k \in K : p \in B_k\}$; then $(B_k)_{k \in J}$ is a countable collection of open sets containing p. We claim it is a neighbourhood base for p. To show this, let V be any neighbourhood of p. Then V includes an open set U which contains p. Hence U includes one of the sets B_k which contains p, i.e. one of the sets B_k with k in J.

63. (1) Let p be any point of E. Then $\{p\}$ forms a neighbourhood base for p. Every base for T must contain all the sets $\{x\}$, and there are uncountably many of these. So (E, T) is first countable but not second countable.

(2) We have seen in Exercise 60(1) that (E, T) is not second countable. The set $\{p\}$ is a neighbourhood base at p; if $q \neq p$ the set $\{q, p\}$ is a neighbourhood base at q. So (E, T) is first countable.

64. (1) If E is countable, $\{\{x\}\}_{x \in E}$ is a countable base for the discrete topology on E.

(2) If E is any set, $\{E, \emptyset\}$ is a countable base for the trivial topology on E.

(3) The base $\{\{2k, 2k + 1\}\}_{k \in \mathbf{N}}$ for the odd-even topology on $\mathbf{N}$ is countable.

(4) Let $E = \{x_0, x_1, x_2, \ldots\}$ and let T be the particular point topology T_{x_0}. Then we see that $\{\{x_0\}, \{x_0, x_1\}, \{x_0, x_2\}, \ldots\}$ is a countable base for T.

(5) Let $E = \{x_0, x_1, x_2, \ldots\}$ and let T be the excluded point topology T_{-x_0}. Then $\{E, \{x_1\}, \{x_2\}, \{x_3\}, \ldots\}$ is a countable base for T.

(6) The set of all open intervals with rational endpoints is a countable base for the usual topology on $\mathbf{R}$.

(7) The set of all intervals of the form $\{x \in \mathbf{R} : x > r\}$ with r rational is a countable base for the right order topology on $\mathbf{R}$.

Chapter 9

ANSWERS FOR CHAPTER 2

65. (a) Suppose f is continuous.

Let U' be a T'-open subset of E'. Let x be any point of $f^{\leftarrow}(U')$. Then $f(x) \in U'$ and so U' is a T'-neighbourhood of $f(x)$. Since f is continuous there is a T-neighbourhood V of x such that $f^{\rightarrow}(V) \subseteq U'$. Then $x \in V \subseteq f^{\leftarrow}(f^{\rightarrow}(V)) \subseteq f^{\leftarrow}(U')$. So $f^{\leftarrow}(U')$ is a T-neighbourhood of x. Hence $f^{\leftarrow}(U')$ is T-open.

(b) Suppose $f^{\leftarrow}(U')$ is T-open for every T'-open subset of E'.

Let F' be any T'-closed subset of E'. Then $C_{E'}(F')$ is T'-open and hence $C_E(f^{\leftarrow}(F')) = f^{\leftarrow}(C_{E'}(F'))$ is T-open. So $f^{\leftarrow}(F')$ is T-closed.

(c) Suppose $f^{\leftarrow}(F')$ is T-closed for every T'-closed subset F' of E'.

Let x be any point of E, V' any T'-neighbourhood of $f(x)$. Then there is a T'-open set U' such that $f(x) \in U' \subseteq V'$. Then $C_{E'}(U')$ is T'-closed and hence $C_E(f^{\leftarrow}(U')) = f^{\leftarrow}(C_{E'}(U'))$ is T-closed. Thus $f^{\leftarrow}(U')$ is T-open and so is a T-neighbourhood of x; since we have $f^{\rightarrow}(f^{\leftarrow}(U')) \subseteq U' \subseteq V'$, it follows that f is continuous.

66. Let U_3 be any T_3-open subset of E_3. Then, since g is continuous, $g^{\leftarrow}(U_3)$ is a T_2-open subset of E_2 and so, since f is continuous, $f^{\leftarrow}(g^{\leftarrow}(U_3))$ is a T_1-open subset of E_1. But $(g \circ f)^{\leftarrow}(U_3) = f^{\leftarrow}(g^{\leftarrow}(U_3))$. So $g \circ f$ is continuous.

67. If there is a mapping g from E' to E such that $g \circ f = I_E$ and $f \circ g = I_{E'}$, then f is a bijection and g is its inverse. The desired result follows at once.

68. (1) Suppose f is a (T, T')-homeomorphism. Then f is (T, T')-continuous.

To show that f is (T, T')-open, let U be any T-open set. Since f^{-1} is (T', T)-continuous, it follows that $(f^{-1})^{\leftarrow}(U) = f^{\rightarrow}(U)$ is T'-open. So f is (T, T')-open.

Similarly f is (T, T')-closed.

(2) Conversely, suppose f is (T, T')-continuous and (T, T')-open.

To show that f is a (T, T')-homeomorphism it is enough to show that f^{-1} is (T', T)-continuous. So let U be any T-open subset of E. Then $(f^{-1})^{\leftarrow}(U) = f^{\rightarrow}(U)$ is T'-open. Thus f^{-1} is (T', T)-continuous.

A similar argument works if f is (T, T')-closed.

69. Let (a, b) be any point of $\mathbf{R}^2$.

Let V' be any neighbourhood of a in $\mathbf{R}$. There is a positive real number r such that $V(a, r) \subseteq V'$. Let $V = V((a, b), r)$. Then we have $f^{\rightarrow}(V) \subseteq V'$. So f is continuous.

The set of points $\{(x, y) \in \mathbf{R}^2 : xy = 1\}$ is closed in $\mathbf{R}^2$. But its image under f, the set $\{x \in \mathbf{R} : x \neq 0\}$, is not closed in $\mathbf{R}$. So f is not closed.

70. (1) Suppose f is (T, T')-continuous.

Let X' be any subset of E'. Then $\mathrm{Cl}_{T'}(X')$ is a T'-closed subset of E' and so $f^{\leftarrow}(\mathrm{Cl}_{T'}(X'))$ is a T-closed subset of E. Since $X' \subseteq \mathrm{Cl}_{T'}(X')$ it follows that $f^{\leftarrow}(X') \subseteq f^{\leftarrow}(\mathrm{Cl}_{T'}(X'))$. Since $\mathrm{Cl}_T(f^{\leftarrow}(X'))$ is the smallest T-closed set including $f^{\leftarrow}(X')$ it follows that we have $\mathrm{Cl}_T(f^{\leftarrow}(X')) \subseteq f^{\leftarrow}(\mathrm{Cl}_{T'}(X'))$.

(2) Conversely, suppose that for every subset X' of E' we have $\mathrm{Cl}_T(f^{\leftarrow}(X')) \subseteq f^{\leftarrow}(\mathrm{Cl}_{T'}(X'))$.

Let F' be any T'-closed subset of E'. Then $f^{\leftarrow}(\mathrm{Cl}_{T'}(F')) = f^{\leftarrow}(F')$ and so $f^{\leftarrow}(F') \supseteq \mathrm{Cl}_T(f^{\leftarrow}(F')) \supseteq f^{\leftarrow}(F')$. Thus $f^{\leftarrow}(F') = \mathrm{Cl}_T(f^{\leftarrow}(F'))$ and hence is T-closed. So f is continuous.

71 (1) Suppose f is (T, T')-open.

Let X be any subset of E. Then, since $\mathrm{Int}_T(X) \subseteq X$ we have

$f^{\rightarrow}(\text{Int}_T(X)) \subseteq f^{\rightarrow}(X)$. Since f is (T, T')-open, $f^{\rightarrow}(\text{Int}_T(X))$ is a T'-open subset of $f^{\rightarrow}(X)$. Hence $f^{\rightarrow}(\text{Int}_T(X)) \subseteq \text{Int}_{T'}(f^{\rightarrow}(X))$.

(2) Conversely, suppose that for every subset X of E we have $f^{\rightarrow}(\text{Int}_T(X)) \subseteq \text{Int}_{T'}(f^{\rightarrow}(X))$.

Let X be any T-open subset of E.

Then we have $f^{\rightarrow}(X) = f^{\rightarrow}(\text{Int}_T(X)) \subseteq \text{Int}_{T'}(f^{\rightarrow}(X)) \subseteq f^{\rightarrow}(X)$. So $\text{Int}_{T'}(f^{\rightarrow}(X)) = f^{\rightarrow}(X)$ and hence $f^{\rightarrow}(X)$ is T'-open. Hence f is (T, T')-open.

72. (1) Suppose f is a (T, T')-homeomorphism.

Let T_0 be any topology on E' such that f is (T, T_0)-continuous.

Let $U_0 \in T_0$. Then $f^{\leftarrow}(U_0) \in T$. Since f^{-1} is (T', T)-continuous, $(f^{-1})^{\leftarrow}(f^{\leftarrow}(U_0)) \in T'$, i.e. $U_0 \in T'$. Thus $T_0 \subseteq T'$. So T' is the finest topology T_0 on E' such that f is (T, T_0)-continuous.

(2) Conversely, suppose that T' is the finest topology T_0 such that f is (T, T_0)-continuous.

To show that f is a (T, T')-homeomorphism, all that remains is to show that f^{-1} is (T', T)-continuous. If this is not the case, then there is a T-open set U such that $(f^{-1})^{\leftarrow}(U) = f^{\rightarrow}(U)$ is not T'-open. Let T_0 be the topology on E' generated by $T' \cup \{f^{\rightarrow}(U)\}$. Then f is (T, T_0)-continuous and so $T_0 \subseteq T'$. So $f^{\rightarrow}(U) \in T'$, which is a contradiction. So f^{-1} is (T', T)-continuous and hence f is a (T, T')-homeomorphism.

73. (1) Suppose (E, T) is a discrete topological space.

Let (E', T') be any topological space, f any mapping from E to E'. Let U' be any T'-open subset of E'. Then we have $f^{\leftarrow}(U') \in \mathbf{P}(E) = T$. So f is (T, T')-continuous.

Conversely, suppose that for every topological space (E', T') every mapping f from E to E' is (T, T')-continuous.

In particular, $I_E : E \to E$ is $(T, \mathbf{P}(E))$-continuous. Thus for every subset X of E we have $(I_E)^{\leftarrow}(X) = X \in T$. So $T = \mathbf{P}(E)$, i.e. T is the discrete topology.

(2) Let (E, T) be a space with the trivial topology, (E', T') any topological space and f a mapping from E' to E.

The only sets in T are E and $\emptyset$. Since $f^{\leftarrow}(E) = E'$ and $f^{\leftarrow}(\emptyset) = \emptyset$ are both in T' it follows that f is (T', T)-continuous.

Conversely, suppose that for every topological space (E', T') every mapping f from E' to E is (T', T)-continuous.

In particular, $I_E : E \to E$ is $(\{E, \emptyset\}, T)$-continuous. Let U be any T-open set. Then $U = (I_E)^{\leftarrow}(U) \in \{E, \emptyset\}$. So $T = \{E, \emptyset\}$.

74. Let $A = \{x \in E : f(x) = g(x)\}$. Another way of expressing this is to say that $A = \{x \in E : (f - g)(x) = 0\} = (f - g)^{\leftarrow}\{0\}$. Since $\{0\}$ is a closed subset of $\mathbf{R}$ and $f - g$ is continuous, it follows that $(f - g)^{\leftarrow}\{0\}$ is T-closed.

Suppose $K = \{x \in E : f(x) = g(x)\} \supseteq D$ where D is dense. Since $\mathrm{Cl}\, D = E$ the smallest closed set which includes D is E. But K is closed. So $K = E$. Thus $f(x) = g(x)$ for all x in E.

Chapter 10

ANSWERS FOR CHAPTER 3

75. Let T_0 be the topology induced on E by the family $(f_i)_{i \in I}$. Let $T(S)$ be the topology generated by S.

Let U_i be any T_i-open subset of E_i. Then, since f_i is (T_0, T_i)-continuous, $f^{\leftarrow}(U_i) \in T_0$. So $S \subseteq T_0$ and hence $T(S) \subseteq T_0$.

Again let U_i be any T_i-open subset of E_i. Since $f^{\leftarrow}(U_i) \in S$ it follows that $f^{\leftarrow}(U_i) \in T(S)$. So all the mappings f_i are $(T(S), T_i)$-continuous. Thus $T(S) \in \mathbf{A}$ and so $T(S) \supseteq T_0$.

So $T(S) = T_0$, as required.

76. (1) Suppose g is (T', T)-continuous. Since each mapping f_i is (T, T_i)-continuous it follows that each mapping $f_i \circ g$ is (T', T_i)-continuous.

(2) Conversely, suppose that each mapping $f_i \circ g$ is (T', T_i)-continuous.

Let U be any T-open subset of E.

Then $U = \bigcup_{k \in K} \bigcap_{j \in J_k} f_j^{\leftarrow}(U_j)$ where each J_k is a finite subset of I and for each index j in J_k the set U_j is T_j-open. We have $g^{\leftarrow}(U) = \bigcup_{k \in K} \bigcap_{j \in J_k} g^{\leftarrow}(f_j^{\leftarrow}(U_j)) = \bigcup_{k \in K} \bigcap_{j \in J_k} (f_j \circ g)^{\leftarrow}(U_j)$, which is T'-open since each set $(f_j \circ g)^{\leftarrow}(U_j)$ is T'-open and all the index sets J_k are finite. So g is (T', T)-continuous.

77. (1) B is T_A-closed $\Longrightarrow C_A(B)$ is T_A-open $\Longrightarrow$ there is a T-open subset U of E such that $C_A(B) = A \cap U$. Then $C_E(U)$ is T-closed and $B = A \cap C_E(U)$.

Conversely, suppose that $B = A \cap F$ where F is T-closed. Then $C_A(B) = A \cap C_E(F)$; since $C_E(F)$ is T-open it follows that $C_A(B)$ is T_A-open. So B is T_A-closed.

(2) Since $A \supseteq B$ and $\mathrm{Cl}_T B \supseteq B$ we have $A \cap \mathrm{Cl}_T B \supseteq B$. Since $\mathrm{Cl}_T B$ is T-closed, $A \cap \mathrm{Cl}_T B$ is T_A-closed. So $A \cap \mathrm{Cl}_T B \supseteq \mathrm{Cl}_{T_A} B$.

Since $\mathrm{Cl}_{T_A} B$ is T_A-closed, there is a T-closed set F such that $\mathrm{Cl}_{T_A} B = A \cap F$. Then $F \supseteq B$ and so $F \supseteq \mathrm{Cl}_T B$. It follows that $\mathrm{Cl}_{T_A} B = A \cap F \supseteq A \cap \mathrm{Cl}_T B$. So $\mathrm{Cl}_{T_A} B = A \cap \mathrm{Cl}_T B$.

(3) $\mathrm{Int}_T B \subseteq B$, so $A \cap \mathrm{Int}_T B = \mathrm{Int}_T B$ and hence $\mathrm{Int}_T B$ is T_A-open. Hence $\mathrm{Int}_T B \subseteq \mathrm{Int}_{T_A} B$.

To show that the inclusion in (3) may be proper, let B be any subset of E which is not T-open; then, if $A = B$ we have $\mathrm{Int}_{T_A} B = B$ but, since B is not T-open, $\mathrm{Int}_T B$ is properly included in B.

78. Since V is a T_D-neighbourhood of a, there is a T-open set U such that $a \in D \cap U \subseteq V$.

Let x be any point of U, W any T-neighbourhood of x. Then $U \cap W$ is a T-neighbourhood of x. Since D is dense we have $x \in \mathrm{Cl}_T D(= E)$; it follows that $(U \cap W) \cap D \neq \emptyset$. Hence $W \cap (D \cap U) \neq \emptyset$; so $W \cap V \neq \emptyset$. Hence $x \in \mathrm{Cl}_T V$. Thus $a \in U \subseteq \mathrm{Cl}_T V$. So $\mathrm{Cl}_T V$ is a T-neighbourhood of a.

79. Since M is T_A-open there exists a T-open set U such that $M = A \cap U$. Similarly, since M is T_B-open there is a T-open set V such that $M = B \cap V$.

Then we have $A \cap U \cap V = B \cap V \cap V = B \cap V = M$ and $B \cap U \cap V = A \cap U \cap U = A \cap U = M$. So $M = M \cup M = [A \cap (U \cap V)] \cup [B \cap (U \cap V)] = (A \cup B) \cap (U \cap V) = U \cap V$, which is T-open.

80. Suppose (E, T) is separable. Let D be a countable dense subset of E. Let U be an open subset of E. Let $D' = D \cap U$; D' is certainly countable.

We claim that $\mathrm{Cl}_{T_U}(D') = U$.

So let x be any point of U, N any T_U-neighbourhood of x. There is a T-open set W such that $x \in U \cap W \subseteq N$. Now $U \cap W$ is a T-neighbourhood of x and $x \in E = \mathrm{Cl}_T D$. So $U \cap W \cap D \neq \emptyset$. Hence $(U \cap W) \cap (U \cap D) \neq \emptyset$, and hence $N \cap D' \neq \emptyset$. Thus $x \in \mathrm{Cl}_{T_U}(D')$. So D' is a countable dense subset of U and so (U, T_U) is separable.

81. According to Exercise 58, $\mathrm{Cl}_{T_p}\{p\} = E$. So $\{p\}$ is a countable dense subset of E. Thus (E, T_p) is separable.

Let D be any subset of A. Then $p \notin D$, so $p \in C_E(D)$ and hence $C_E(D)$ is T_p-open. So D is T_p-closed. Then $\mathrm{Cl}_{(T_p)_A} D = A \cap \mathrm{Cl}_{T_p} D = A \cap D = D$. If D is countable, its closure $\mathrm{Cl}_{(T_p)_A} D$ is also countable and hence is not equal to A, which is uncountable.

So A has no countable dense subset.

82. Let U be any $(\prod T_i)$-open subset of $\prod E_i$.

Then $U = \bigcup_{k \in K} B_k$, where for each index k in K the set $B_k = \prod_{i \in I} X_{ik}$, and for each i in I we have $X_{ik} \in T_i$ while $I_k = \{i \in I : X_{ik} \neq E_i\}$ is finite. Then $\bigcap_{k \in K} I_k$ is finite. If $i_0 \notin \bigcap_{k \in K} I_k$ there is an index k_0 such that $X_{i_0 k_0} = E_{i_0}$. Then $\pi_{i_0}^{\rightarrow}(U) \supseteq \pi_{i_0}^{\rightarrow}(B_{k_0}) = \pi_{i_0}^{\rightarrow}(\prod_{i \in I} X_{ik_0}) = E_{i_0}$.

83. Let $x \in \mathrm{Cl}_T(\prod A_i)$.

Let j be any index in I, V_j any T_j-neighbourhood of $\pi_j(x)$ in E_j. Then V_j includes a T_j-open set U_j which contains $\pi_j(x)$.

Let $U = \prod X_i$ where $X_j = U_j$ and $X_i = E_i$ for $i \neq j$. Then U is a T-open set which contains x. So $U \cap \prod A_i \neq \emptyset$. If $y \in U \cap \prod A_i$ we have $\pi_j(y) \in \pi_j(U) \cap A_j = U_j \cap A_j$. So $U_j \cap A_j$ is non-empty and hence $\pi_j(x) \in \mathrm{Cl}_{T_j} A_j$. Thus $\mathrm{Cl}_T(\prod A_j) \subseteq \prod \mathrm{Cl}_{T_j} A_j$.

Conversely, suppose $x \in \prod \mathrm{Cl}_{T_j} A_j$.

Let V be any T-neighbourhood of x. Then V includes a T-open set containing x and hence a basic T-open set containing x, say U. Let $U = \prod_{i \in I} U_i$ where for each index i in I the set U_i is T_i-open and further $U_i = E_i$ for all indices i not in a certain finite subset J of I. For each index j in J, U_j is a T_j-neighbourhood of $\pi_j(x)$, which belongs to $\mathrm{Cl}_{T_j} A_j$. Hence, for each index j in J, we have $U_j \cap A_j \neq \emptyset$. For $j \notin J$ we have $U_j \cap A_j = E_j \cap A_j = A_j \neq \emptyset$. Hence $U \cap \prod A_j = \prod (U_j \cap A_j) \neq \emptyset$. So $x \in \mathrm{Cl}_T(\prod A_j)$. Hence $\prod \mathrm{Cl}_{T_j} A_j \subseteq \mathrm{Cl}_T(\prod A_j)$.

84. (1) Let $E = \prod_{i \in I} E_i$, $T = \prod_{i \in I} T_i$. For each index i in I, let $D_i = \{p_n^i : n \in \mathbf{N}\}$ be a countable dense subset of E_i.

Let D be the set of all points t of E such that $\pi_i(t) \in D_i$ for all i in I and there is an index $i(t)$ in I and a natural number m such that $\pi_i(t) = p_m^i$ for all $i > i(t)$. D is a countable subset of E. We shall show that D is dense in E.

Let x be any point of E, V any neighbourhood of x; then V includes a T-open set, and hence a basic T-open set U which contains x. Say $U = \prod_{i \in I} U_i$ where $U_i \in T_i$ for all indices i in I and $U_i = E_i$ for all but finitely many indices i. Thus there is an index i_0 such that $U_i = E_i$ for all $i > i_0$.

If $i \leq i_0$ then, since D_i is a dense subset of E_i, there is a point of D_i in U_i, say $p_{n_i}^i$. Let t be the point of D such that $\pi_i(t) = p_{n_i}^i$ for $i \leq i_0$ and $\pi_i(t) = p_0^i$ for all $i > i_0$. Then $t \in D \cap U \subseteq D \cap V$; so $x \in \mathrm{Cl}_T D$, i.e. D is a countable dense subset of E.

(2) Suppose that each space (E_i, T_i) has a countable base B_i for its topology T_i.

Then the collection of subsets $\prod U_i$, where for each i in I we have either $U_i \in B_i$ or $U_i = E_i$ and $U_i = E_i$ for all but a finite number of indices i in I, is countable and is a base for the product topology $\prod T_i$.

85. (1) Suppose f_I is injective; let x, y be distinct points of E. Then $f_I(x) \neq f_I(y)$. So there is at least one index i in I for which $\pi_i f_I(x) \neq \pi_i f_I(y)$, i.e. $f_i(x) \neq f_i(y)$. Thus the family $(f_i)_{i \in I}$ distinguishes points.

Conversely, suppose $(f_i)_{i \in I}$ distinguishes points. Then if x and y are distinct points of E there is an index i in I such that $f_i(x) \neq f_i(y)$ and hence $f_I(x) \neq f_I(y)$. Thus f is injective.

(2) Suppose $(f_i)_{i \in I}$ distinguishes points and closed sets.

Let U be a T-open subset of E. If $t = f_I(x)$ is any point of $(f_I)^{\rightarrow}(U)$ there is an index i in I such that we have $f_i(x) \notin \mathrm{Cl}_{T_i}((f_i)^{\rightarrow}(C_E(U))) = F_i$, say.

Let $U_i = C_{E_i}(F_i)$ and for $j \neq i$ let $U_j = E_j$. Then $X = \prod_{j \in I} U_j$ is open in the product topology $\prod_{j \in I} T_j$ and $t \in X \cap (f_I)^{\rightarrow}(U)$ which is open in the subspace topology T_I on $(f_I)^{\rightarrow}(U)$. Thus $(f_I)^{\rightarrow}(U)$ is a neighbourhood of each of its points in this subspace topology and hence is open in the subspace topology. So f_I is (T, T_I)-open.

86. For each point p of E let f_p be the mapping from E to X given by

$$f_p(t) = \begin{cases} c & \text{if } t \neq p \\ b & \text{if } t = p. \end{cases}$$

To show that f_p is (T, T_0)-continuous we must check that the set $(f_p)^{\leftarrow}(\{a\})$ is T-open. But this is indeed the case since $(f_p)^{\leftarrow}(\{a\}) = \emptyset$.

For each T-open subset U of E, let f_U be the mapping from E to X given by

$$f_U(t) = \begin{cases} a & \text{if } t \in U \\ b & \text{if } t \notin U. \end{cases}$$

Since $(f_U)^{\leftarrow}(\{a\}) = U$, which is T-open, f_U is (T, T_0)-continuous.

Let $I = X \cup T$ and consider the family $(f_i)_{i \in I}$. This clearly distinguishes points since if s and t are distinct points of E we have $f_s(s) = b$ but $f_s(t) = c$. So f_I is an injective mapping.

To show that the family $(f_i)_{i \in I}$ distinguishes points and closed sets let F be any T-closed subset of E, x any point not in F. Let $U = C_E(F)$. Then $f_U(x) = a$ but for every point t in F we have $f_U(t) = b$. So $f^{\rightarrow}(F) = \{b\}$ and $\operatorname{Cl}_{T_U}(f^{\rightarrow}(F))$, the smallest T_0-closed subset which includes $f^{\rightarrow}(F)$ is $\{b, c\}$, which does not contain the point a. So f_I is (T, T_I)-open. Hence f_I is a (T, T_I)-homeomorphism.

87. Let $T' = \{U \in \mathbf{P}(E) : g_i^{\leftarrow}(U) \in T_i \text{ for all } i \text{ in } I\}$. Then T' is a topology on E and each of the mappings g_i is (T_i, T')-continuous. Hence $T' \subseteq T$, the coinduced topology.

On the other hand if V is any T-open set then, since each mapping g_i is (T_i, T)-continuous, each set $g_i^{\leftarrow}(V)$ is T_i-open. So $V \in T'$. Hence $T \subseteq T'$.

88. (1) Suppose f is (T, T')-continuous.

Since each mapping g_i is (T_i, T)-continuous it follows that each mapping $f \circ g_i$ is (T_i, T')-continuous.

(2) Conversely, suppose each mapping $f \circ g_i$ is (T_i, T')-continuous.

Let U' be any T'-open set. For each index i in I we have $(f \circ g_i)^{\leftarrow}(U') \in T_i$, i.e. $g_i^{\leftarrow}(f^{\leftarrow}(U')) \in T_i$. So $f^{\leftarrow}(U') \in T$. Thus f is (T, T')-continuous.

89. (1) Suppose f^* is a $(T/R_f, (T')_B)$-homeomorphism. Let U be a T-open subset of E which is R_f-saturated, say $U = \bigcup_{x \in X} R_f(x)$ for some set X. Then $\eta^{\leftarrow}(\eta^{\rightarrow}(U)) = U$ and so $\eta^{\rightarrow}(U)$ is T/R_f-open. Since

f^* is a $(T/R_f, (T')_B)$-homeomorphism, it is $(T/R_f, (T')_B)$-open. So $(f^*)^{\rightarrow}(\eta^{\rightarrow}(U))$ is $(T')_B$-open. But $f^* \circ \eta = f$; so we have $f^{\rightarrow}(U) = (f^*)^{\rightarrow}(\eta^{\rightarrow}(U)) \in (T')_B$.

(2) Conversely, suppose that for every T-open subset U of E which is R_f-saturated we have $f^{\rightarrow}(U) \in (T')_B$.

Let W be any (T/R_f)-open set. Then $\eta^{\leftarrow}(W)$ is T-open and R_f-saturated. Hence we have $f^{\rightarrow}(\eta^{\leftarrow}(W)) \in (T')_B$. So it follows that $(f^*)^{\rightarrow}(\eta^{\rightarrow}(\eta^{\leftarrow}(W))) \in (T')_B$, i.e. $(f^*)^{\rightarrow}(W) \in (T')_B$.

So f^* is $(T/R_f, (T')_B)$-open and hence is a $(T/R_f, (T')_B)$-homeomorphism.

90. (1) Suppose R is open.

Let U be a T-open set. Then, by hypothesis, $\eta^{\rightarrow}(U)$ is T/R-open. Hence $\eta^{\leftarrow}(\eta^{\rightarrow}(U))$ is T-open. But $\eta^{\leftarrow}(\eta^{\rightarrow}(U))$ is the saturate of U.

(2) Conversely, suppose that the saturate of every T-open set is T-open.

Let U be any T-open subset of E. Then, by hypothesis, $\eta^{\leftarrow}(\eta^{\rightarrow}(U))$ is T-open. So $\eta^{\rightarrow}(U)$ is T/R-open. Hence η is $(T, T/R)$-open, i.e. R is an open relation.

91. (1) $\Longrightarrow$ (2). Suppose R is an open relation.

Let A be an R-saturated subset of E, so that $\eta^{\leftarrow}(\eta^{\rightarrow}(A)) = A$. Let $U = \operatorname{Int}_T A$. Since U is T-open and R is open, $\eta^{\rightarrow}(U)$ is T/R-open and hence $\eta^{\leftarrow}(\eta^{\rightarrow}(U))$ is T-open. Since $U \subseteq A$ we have $\eta^{\leftarrow}(\eta^{\rightarrow}(U)) \subseteq \eta^{\leftarrow}(\eta^{\rightarrow}(A)) = A$. So $\eta^{\leftarrow}(\eta^{\rightarrow}(U)) \subseteq U$, the largest T-open subset included in A. But we always have $U \subseteq \eta^{\leftarrow}(\eta^{\rightarrow}(U))$.

So $U = \eta^{\leftarrow}(\eta^{\rightarrow}(U))$, i.e. U is R-saturated.

(2) $\Longrightarrow$ (3). Suppose that the interior of every R-saturated set is R-saturated.

Let B be any R-saturated subset of E, i.e. let $\eta^{\leftarrow}(\eta^{\rightarrow}(B)) = B$. Since B is the union of the R-classes of all members of B we see that $C_E(B)$ is the union of the R-classes of all elements of E not in B. Thus $C_E(B)$ is R-saturated. Hence, by hypothesis, it follows that $\operatorname{Int} C_E(B) = C_E(\operatorname{Cl} B)$ is R-saturated.

So $\operatorname{Cl} B$ is R-saturated.

(3) $\Longrightarrow$ (1). Suppose the closure of every R-saturated set is R-saturated.

Let U be any T-open subset of E. Let $U' = \eta^{\leftarrow}(\eta^{\rightarrow}(U))$, which is

R-saturated. Then $C_E(U')$ is R-saturated and hence, by hypothesis, $\mathrm{Cl}\, C_E(U') = C_E(\mathrm{Int}\, U')$ is R-saturated. It follows that $\mathrm{Int}\, U'$ is R-saturated. Since $U \subseteq U'$ and U is T-open, we have $U \subseteq \mathrm{Int}\, U'$. Hence $U' = \eta^{\leftarrow}(\eta^{\rightarrow}(U)) \subseteq \eta^{\leftarrow}(\eta^{\rightarrow}(\mathrm{Int}\, U'))$. So $\mathrm{Int}\, U' = U'$ and hence U' is open.

Hence $\eta^{\rightarrow}(U)$ is (T/R)-open; so R is an open relation.

92. (1) Suppose R is closed.

Let C be an R-class, U a T-open set such that $U \supseteq C$. Since $C_E(U)$ is T-closed, $\eta^{\rightarrow}(C_E(U))$ is (T/R)-closed. Hence $C_{E/R}(\eta^{\rightarrow}(C_E(U)))$ is (T/R)-open. Thus it follows that $W = \eta^{\leftarrow}(C_{E/R}(\eta^{\rightarrow}(C_E(U))))$ is T-open and R-saturated.

We have $C \subseteq W$ and $W \subseteq U$.

(2) Conversely, suppose that for every R-class C and every T-open set U including C there exists a saturated T-open set W such that $C \subseteq W \subseteq U$.

Let F be a T-closed subset of E. Let $G = \eta^{\leftarrow}(\eta^{\rightarrow}(F))$; we shall show that G is T-closed, from which it follows that $\eta^{\rightarrow}(F)$ is (T/R)-closed. Let x be any point of $C_E(G)$; let $C = R(x)$. Then $C \cap G = \emptyset$ (since G is a union of R-classes which does not contain x). So $C \cap F = \emptyset$. Thus $C_E(F)$ is a T-open set which includes C. By hypothesis there is a saturated T-open set W such that $C \subseteq W \subseteq C_E(F)$. Then $W \cap G = \eta^{\leftarrow}(\eta^{\rightarrow}(W)) \cap \eta^{\leftarrow}(\eta^{\rightarrow}(F)) = \emptyset$. So $W \subseteq C_E(G)$.

Hence $C_E(G)$ is a T-neighbourhood of each of its points x and hence is T-open. So G is T-closed, as required.

93. Let C be the R-class of $(0,0)$, which consists of $(0,0)$ alone.

Let $U = \{(x,y) \in \mathbf{R}^2 : x^2 + y^2 < 1 \text{ and } y \geq 0\}$; U is an open subset of E and $U \supseteq C$. Any open set W included in U must contain points (x,y) with $y \neq 0$. But the R-class of such a point is not included in U. So R is not closed.

94. Let $\mathbf{K}$ be the set of R-classes C such that $C \subseteq U$. Then certainly $\bigcup \mathbf{K} \subseteq \bigcup_{C \in \mathbf{K}} U_C$.

Conversely, suppose $x \in \bigcup_{C \in \mathbf{K}} U_C$, say $x \in U_{C_1}$. U_{C_1} is a union of R-classes; so x belongs to one of these classes, which is of course in $\mathbf{K}$ since $U_{C_1} \subseteq U$. So $x \in \bigcup \mathbf{K}$. Thus $\bigcup_{C \in \mathbf{K}} U_C \subseteq \bigcup \mathbf{K}$.

95. Since X is saturated we have $\eta^{\leftarrow}(\eta^{\rightarrow}(X)) = X$. Suppose $\bar{A} \in (T/R)_{\eta^{\rightarrow}(X)}$. Then there is a set $\bar{B}$ in T/R such that $\bar{A} = \eta^{\rightarrow}(X) \cap \bar{B}$. Then $\eta^{\leftarrow}(\bar{A}) = \eta^{\leftarrow}(\eta^{\rightarrow}(X)) \cap \eta^{\leftarrow}(\bar{B}) = X \cap \eta^{\leftarrow}(\bar{B})$ and $X \cap \eta^{\leftarrow}(\bar{B}) \in T_X$ since $\eta^{\leftarrow}(\bar{B}) \in T$. So $\bar{A} \in (T_X)/(R \,|\, (X \times X))$.

Conversely, let $\bar{A}$ be a set in $(T_X)/(R \,|\, (X \times X))$.

Then $\eta^{\leftarrow}(\bar{A}) \in T_X$ and so there is a T-open subset U such that $\eta^{\leftarrow}(\bar{A}) = U \cap X$. Let $\mathbf{K}$ be the set of R-classes C such that $C \subseteq U$. Then $B = \bigcup_{C \in \mathbf{K}} C = \bigcup_{C \in \mathbf{K}} U_C$ and hence is T-open. Of course $B \supseteq \eta^{\leftarrow}(\bar{A})$ since for every element $\bar{a}$ of $\bar{A}$ we have $\eta^{\leftarrow}(\bar{a}) \subseteq U$. Hence $X \cap B \supseteq \eta^{\leftarrow}(\bar{A}) = X \cap U \supseteq X \cap B$. So $\eta^{\leftarrow}(\bar{A}) = X \cap B$. Now $B = \eta^{\leftarrow}(\eta^{\rightarrow}(B))$ is T-open; so $\eta^{\rightarrow}(B)$ is (T/R)-open. Now $\bar{A} = \eta^{\rightarrow}(\eta^{\leftarrow}(\bar{A})) = \eta^{\rightarrow}(X \cap B) = \eta^{\rightarrow}(X) \cap \eta^{\rightarrow}(B)$. Thus $\bar{A} \in (T/R)_{\eta^{\rightarrow}(X)}$.

Chapter 11

ANSWERS FOR CHAPTER 4

96. (1) Let $F = \{X \in \mathbf{P}(E) : X \supseteq A\}$.

If $X \in F$ and $Y \supseteq X$ then, since $X \supseteq A$ and $Y \supseteq X$, we have $Y \supseteq A$, i.e. $Y \in F$.

If $X_1, X_2 \in F$ then $X_1 \supseteq A$ and $X_2 \supseteq A$; so $X_1 \cap X_2 \supseteq A$, i.e. $X_1 \cap X_2 \in F$.

If $X \in F$ then $X \supseteq A \neq \emptyset$, so $X \neq \emptyset$.

Thus F is a filter on E.

(2) $\{A\} \subseteq F$ and if $X \in F$ then $X \supseteq A$. So $\{A\}$ is a base for the filter F.

97. Routine checking.

98. Routine checking.

99. (1) Suppose $X \in F \cap G$. Then $X = X \cup X$, $X \in F$ and $X \in G$.

(2) Conversely, suppose $X = P \cup Q$ where $P \in F$ and $Q \in G$. Then $X \supseteq P$, so $X \in F$, and $X \supseteq Q$, so $X \in G$.

100. (1) Suppose B is a base for a filter F on E.

Let $(X_i)_{1 \le i \le n}$ be a finite family of sets in B. Since $B \subseteq F$ it follows that $X_1 \cap \ldots \cap X_n \in F$ and so $X_1 \cap \ldots \cap X_n$ includes a set in B.

Since F is non-empty and every set in F includes a set in B, it follows that B is non-empty.

Since $\emptyset \notin F$ and $B \subseteq F$, we have $\emptyset \notin B$.

(2) Conversely, suppose the conditions are satisfied.

Let $F = \{X \in \mathbf{P}(E) : X$ includes a set in the collection $B\}$. Then F is a filter on E with base B.

101. Let Φ be the filter generated by $F \cup G$.

Let S be the set of intersections of all finite families of sets from $F \cup G$. Then $X \in \Phi \Longrightarrow X \supseteq$ a set in S. Every set in S has the form $P \cap Q$ where $P \in F$ and $Q \in G$. If $X \supseteq P \cap Q$ where $P \in F$ and $Q \in G$, then it follows that we have $X = X \cup X = X \cup (P \cap Q) = (X \cup P) \cap (X \cup Q)$. Since $X \cup P \supseteq P$ and $X \cup Q \supseteq Q$ we have $X \cup P \in F$ and $X \cup Q \in G$. So $X \in S$. Thus $\Phi = S$, as required.

102. Let $\mathbf{F}$ be a set of filters on E totally ordered by inclusion.

Let $\mathbf{A}$ be the union of $\mathbf{F}$. Let $(X_i)_{i \in I}$ be a finite family of sets in $\mathbf{A}$. For each index i in I there is a filter F_i in $\mathbf{F}$ such that $X_i \in F_i$. Since $\mathbf{F}$ is ($\subseteq$)-totally ordered, there is an index j in I such that $X_i \in F_j$ for all i in I. Hence $\bigcap_{i \in I} X_i \neq \emptyset$. So $\mathbf{A}$ generates a filter F on E which is clearly the ($\subseteq$)-supremum of $\mathbf{F}$.

103. Suppose $A \notin F$, $B \notin F$.

Let $F' = \{X \in \mathbf{P}(E) : A \cup X \in F\}$. Then F' is a filter on E (notice that $\emptyset \notin F'$ since $A \notin F$). Clearly $F' \supseteq F$ and $B \in F'$ although $B \notin F$. So $F' \supset F$, which contradicts the fact that F is an ultrafilter.

104. Let F be the filter generated by $\mathbf{A}$, F' any ultrafilter which includes F. Of course $F' \supseteq \mathbf{A}$. Let X be any set in F'. Then $C_E(X) \notin \mathbf{A}$, for if $C_E(X) \in \mathbf{A}$ then $C_E(X) \in F'$ and $X \cap C_E(X) = \emptyset \in F'$. This is a contradiction since F' is a filter. Hence $X \in \mathbf{A}$ and so $F' \subseteq \mathbf{A}$. So $\mathbf{A} = F'$, an ultrafilter.

105. Let F be an ultrafilter.

Suppose $\bigcap F$ contains two distinct points a and b. Let F' be the filter generated by $F \cup \{a\}$; clearly $F' \supseteq F$ and $\{a\} \in F'$. But $\{a\}$ does not belong to F since it does not contain b. Thus $F' \supset F$, which is a contradiction since F is an ultrafilter.

So $\cap F$ cannot contain more than one point.

If $\cap F = \{a\}$ then $F \subseteq G$, the filter of all sets which include $\{a\}$. Hence $F = G$, since F is an ultrafilter.

106. (1) Suppose F_A is a filter on A. Then all the sets of F_A are non-empty.

(2) Conversely, suppose all the sets of F_A are non-empty.

If $X \cap A \in F_A$ and Y' is a subset of A such that $Y' \supseteq X \cap A$ we have $Y' = Y' \cup (X \cap A) = (Y' \cup X) \cap (Y' \cup A) = (Y' \cup X) \cap A \in F_A$ since $Y' \cup X \in F$ (because $X \in F$ and $Y' \cup X \supseteq X$).

If $(X_i \cap A)_{i \in I}$ is a finite family of sets in F_A then $\cap (X_i \cap A) = (\cap X_i) \cap A \in F_A$ since $\cap X_i \in F$.

By hypothesis all the sets in F_A are non-empty.

So F_A is a filter on A.

(3) If $A \in F$ then $X \cap A \neq \emptyset$ for all X in F. So F_A is a filter on A.

(4) Conversely, suppose F is an ultrafilter and $X \cap A \neq \emptyset$ for all X in F. If $A \notin F$ then $F \cup \{A\}$ generates a filter which properly includes F; this is impossible since F is an ultrafilter. So $A \in F$.

(5) Suppose F is an ultrafilter and F_A is a filter on A. If F_A is not an ultrafilter on A there is a filter F' on A properly including F_A. Let Y' be a subset of A which belongs to F' but not to F_A. Then $F \cup \{Y'\}$ is a filter on E which properly includes F. This is impossible. So F_A is an ultrafilter.

107. Let X be a set which does not belong to F.

Then for each set M in F we cannot have $M \subseteq X$ and hence we must have $M \cap C_E(X) \neq \emptyset$. So $F \cup \{C_E(X)\}$ generates a filter on E, which is included in some ultrafilter F_X. Since $C_E(X) \in F_X$ we must have $X \notin F_X$. Thus X does not belong to the intersection of the set of all ultrafilters which include F. Hence this intersection is just the filter F itself.

108. (1) Since $B \neq \emptyset$ we have $f^{\to}(B) \neq \emptyset$.

Since $\emptyset \notin B$ we have $\emptyset \notin f^{\to}(B)$.

Let X_1', X_2' be sets in $f^{\to}(B)$; then there are sets X_1, X_2 in B such that $X_1' = f^{\to}(X_1)$ and $X_2' = f^{\to}(X_2)$. Since X_1, $X_2 \in B$ there is a set W in B such that $W \subseteq X_1 \cap X_2$. Then we must have $f^{\to}(W) \subseteq f^{\to}(X_1 \cap X_2) \subseteq f^{\to}(X_1) \cap f^{\to}(X_2) = X_1' \cap X_2'$.

So $f^\rightarrow(B)$ is a filter base on E'.

(2) Suppose B is base for an ultrafilter F on E.

Let F' be the filter on E' generated by $f^\rightarrow(B)$. Let X' be any subset of E'. If $f^\leftarrow(X') \in F$ then $f^\leftarrow(X')$ includes a set W in B. Then $X' \supseteq f^\rightarrow(f^\leftarrow(X')) \supseteq f^\rightarrow(W)$ and hence $X' \in F'$. If $f^\leftarrow(X') \notin F$ then, since F is an ultrafilter on E, $C_E(f^\leftarrow(X')) \in F$ and hence $C_E(f^\leftarrow(X')) = f^\leftarrow(C_{E'}(X'))$ includes a set W in B. Then $C'_E(X') \supseteq f^\rightarrow(W)$ and hence $C'_E(X') \in F'$. So F' is an ultrafilter on E'.

109. (1) Suppose $f^\leftarrow(B')$ is base for a filter on E.

Then for every set X' in B' we have $f^\leftarrow(X') \neq \emptyset$ and hence $X' \cap f^\rightarrow(E) \neq \emptyset$.

(2) Conversely, suppose that $X' \cap f^\rightarrow(E) \neq \emptyset$ for all X' in B'.

Then $\emptyset \notin f^\leftarrow(B')$ and $f^\leftarrow(B') \neq \emptyset$ (since $B' \neq \emptyset$). Suppose X_1 and X_2 are sets in $f^\leftarrow(B')$. Then there are sets X'_1, X'_2 in B' such that $X_1 = f^\leftarrow(X'_1)$ and $X_2 = f^\leftarrow(X'_2)$. There is a set W' in B' such that $W' \subseteq X'_1 \cap X'_2$. Then $f^\leftarrow(W') \subseteq f^\leftarrow(X'_1 \cap X'_2) = f^\leftarrow(X'_1) \cap f^\leftarrow(X'_2) = X_1 \cap X_2$. So $f^\leftarrow(B')$ is a filter base on E.

110. (1) Suppose T is finer than T'.

Let F be a filter which is T-convergent to a. Then $F \supseteq V_T(a)$, the T-neighbourhood filter of a. Since T is finer than T', every T'-neighbourhood of a is a T-neighbourhood. So $F \supseteq V_{T'}(a)$, the T'-neighbourhood filter of a, and hence F is T'-convergent to a.

(2) Conversely, suppose that every filter on E which is T-convergent to a is also T'-convergent to a.

Let U' be a T'-open set, p any point of U'. Then $U' \in V_{T'}(p)$. Since $V_T(p)$ is T-convergent to p, it follows from our hypothesis that it is T'-convergent to p. Thus $V_T(p) \supseteq V_{T'}(p)$ and in particular $U' \in V_T(p)$. Thus U' is a T-neighbourhood of each of its points and hence is T-open. So $T' \subseteq T$, i.e. T is finer than T'.

111. Let F be a filter on E which converges to p.

Let V' be any T'-neighbourhood of $f(p)$. Since f is (T, T')-continuous at p, there is a T-neighbourhood V of p such that $f^\rightarrow(V) \subseteq V'$. Since F converges to p we have $V \in F$ and so $f^\rightarrow(V) \in f^\rightarrow(F)$. So V' belongs to the filter based on $f^\rightarrow(F)$. Hence $f^\rightarrow(F)$ converges to $f(p)$.

112. Let F be the filter of which B is a base.

Then, according to the definition of the adherence of a filter base, $\operatorname{Adh} B = \operatorname{Adh} F = \bigcap_{X \in F} \operatorname{Cl} X \subseteq \bigcap_{X \in B} \operatorname{Cl} X$.

Let X_0 be any set in F. Then there is a set B_0 in B such that $X_0 \supseteq B_0$ and so $\operatorname{Cl} X_0 \supseteq \operatorname{Cl} B_0 \supseteq \bigcap_{X \in B} \operatorname{Cl} X$. Thus we have $\bigcap_{X \in F} \operatorname{Cl} X \supseteq \bigcap_{X \in B} \operatorname{Cl} X$. Hence $\bigcap_{X \in F} \operatorname{Cl} X = \bigcap_{X \in B} \operatorname{Cl} X$.

113. (1) Suppose x is adherent to A.

Then every T-neighbourhood V of x meets A. It follows that $V_T(x) \cup \{A\}$ generates a filter which contains A and is T-convergent to x.

(2) Conversely, suppose there is a filter F such that $A \in F$ and F is T-convergent to x.

Let V be any T-neighbourhood of x. Then $V \in F$, and since $A \in F$ it follows that $A \cap V \neq \emptyset$. So x is adherent to A.

114. (1) Suppose x is a limit point of B. Thus the filter F based on B converges to x. Let V be any set in N. Then $V \in F$. Hence V includes a set in B.

Conversely, suppose every set in N includes a set in B.

Let V be any T-neighbourhood of x. Then V includes a set in N and hence a set in B. So V belongs to the filter F based on B. Hence F (and so B) converges to x.

(2) Suppose x is an adherent point of B.

Then x is adherent to every set in the filter generated by B. So every T-neighbourhood of x meets every set in that filter. Since every set in N is a T-neighbourhood of x and every set in B belongs to the filter, it follows that every set in N meets every set in B.

Conversely, suppose every set in N meets every set in B.

Let V be any T-neighbourhood of x, X any set in the filter based on B. Then V includes a set W in N and X includes a set Y in B. Since $W \cap Y \neq \emptyset$ it follows that $V \cap X \neq \emptyset$. So x is adherent to B.

115. (1) Suppose x is adherent to F.

Then every set in $V_T(x)$ meets every set in F. Hence $F \cup V_T(x)$ generates a filter F' on E. Clearly $F' \supseteq F$ and F' converges to x.

(2) Conversely, suppose $F \subseteq F'$ where F' is a filter which converges

to x.

Then every T-neighbourhood of x belongs to F'. Since every set in F also belongs to F', it follows that every set in F meets every T-neighbourhood of x. So x is adherent to F.

116. If x is a limit point of a filter F then $V_T(x) \subseteq F$ and so every T-neighbourhood of x meets every set in F. So x is adherent to F.

117. If x is adherent to an ultrafilter F there is a filter F' such that $F' \supseteq F$ and F' converges to x. But, since F is an ultrafilter, $F' \supseteq F \Longrightarrow F' = F$. So F converges to x.

118. (1) Suppose f converges to x relative to F.

Let V_i be a T_i-neighbourhood of $\pi_i(x)$. Then V_i includes a T_i-open set U_i containing $\pi_i(x)$. Let $V = \prod_{j \in I} Y_j$ where $Y_j = E_j$ for $j \neq i$ and $Y_i = U_i$. Then V is a T-open subset of E containing x, hence a T-neighbourhood of x. Hence there is a set X in F such that $V \supseteq f^\rightarrow(X)$. Then $U_i = (\pi_i)^\rightarrow(V) \supseteq (\pi_i)^\rightarrow(f^\rightarrow(X)) = (\pi_i \circ f)^\rightarrow(X)$. So $\pi_i \circ f$ converges to $\pi_i(x)$ relative to F.

(2) Conversely, suppose each $\pi_i \circ f$ converges to $\pi_i(x)$ relative to the filter F.

Let V be any T-neighbourhood of x. Then there is a family $(U_i)_{i \in I}$ where each U_i is T_i-open, $U_i = E_i$ for all i not in a certain finite subset J of I and $x \in \prod_{i \in I} U_i \subseteq V$. For each j in J, U_j is a T_j-neighbourhood of $\pi_j(x)$. Then there is a set X_j in F such that $(\pi_j \circ f)^\rightarrow(X_j) \subseteq U_j$. Let $X = \bigcap_{j \in J} X_j$; then $X \in F$ and $f^\rightarrow(X) \subseteq V$. So f is convergent to x relative to F.

119. (1) Suppose x' is a limit point of f relative to F.

Let V' be a T'-neighbourhood of x'. Then there is a set X in F such that $f^\rightarrow(X) \subseteq V'$. Then $f^\leftarrow(V') \supseteq f^\leftarrow(f^\rightarrow(X)) \supseteq X$ and so $f^\leftarrow(V') \in F$.

Conversely, suppose that for every neighbourhood V' of x' we have $f^\leftarrow(V') \in F$.

Then for every neighbourhood V' of x' we have $V' \supseteq f^\rightarrow(f^\leftarrow(V'))$. So x' is a limit point of f relative to F.

(2) Suppose x' is an adherent point of f relative to F.

Let V' be a T'-neighbourhood of x', X any set in F. Then

$V' \cap f^{\rightarrow}(X) \neq \emptyset$ (since x is adherent to $f^{\rightarrow}(F)$).

The converse is clear.

120. (1) Suppose f is (T, T')-continuous at x.

Let V' be a T'-neighbourhood of $f(x)$. Then there is a T-neighbourhood V of x such that $V' \supseteq f^{\rightarrow}(V)$. So $f(x)$ is a (T, T')-limit point of f at x.

(2) Conversely, suppose $f(x)$ is a (T, T')-limit point of f at x.

Let V' be any T'-neighbourhood of $f(x)$. Then there is a T-neighbourhood V of x such that $f^{\rightarrow}(V) \subseteq V'$.

So f is (T, T')-continuous at x.

121. Let R be the relation on $\mathbf{A}$ given by $(X, Y) \in R$ if and only if $X \supseteq Y$. Then $(\mathbf{A}, R)$ is a directed set.

Let $D' = D \times \mathbf{A}$ and let R' be the order relation on D' given by $((d, A), (d_1, A_1)) \in R'$ if and only if $d \le d_1$ and $(A, A_1) \in R$.

Let $\bar{D} = \{(d, A) \in D' : \nu(d) \in A\}$; let $\bar{R}$ be the restriction of R' to $\bar{D} \times \bar{D}$. We claim that $(\bar{D}, \bar{R})$ is a directed set. To see this, let (d_1, A_1), (d_2, A_2) be elements of $\bar{D}$. There exists an element A of $\mathbf{A}$ such that $A \subseteq A_1 \cap A_2$, whence $(A_1, A) \in R$, $(A_2, A) \in R$. Since ν is frequently in A there is an element d of D such that $d \ge d_1$, $d \ge d_2$ and $\nu(d) \in A$. Then $(d, A) \in \bar{D}$ and is $\bar{R}$-greater than (d_1, A_1) and (d_2, A_2).

Let φ be the mapping from $\bar{D}$ to D given by $\varphi(d, A) = d$ for all (d, A) in $\bar{D}$. Let $\nu' = \nu \circ \varphi$; we shall show that ν' is a subnet of ν.

To this end, let d be any element of D. Let A be any element of $\mathbf{A}$. Since ν is frequently in A there is an element d_1 of D such that $d_1 \ge d$ and $\nu(d_1) \in A$. Now let (d_2, A_2) be any element of $\bar{D}$ which is $\bar{R}$-greater than (d_1, A). Then $\varphi(d_2, A_2) = d_2 \ge d_1 \ge d$. So ν' is a subnet of ν.

Next we show that ν' is eventually in every set in $\mathbf{A}$.

So let A be any element of $\mathbf{A}$. By hypothesis there is an element d of D such that $\nu(d) \in A$. Let (d_1, A_1) be any element of $\bar{D}$ which is $\bar{R}$-greater than (d, A). Then $\nu'(d_1, A_1) = \nu(d_1) \in A_1 \subseteq A$. So ν' is eventually in A.

122. (1) Suppose a is an adherent point of ν.

Then ν is frequently in every T-neighbourhood of a. The set $V_T(a)$ satisfies the conditions on $\mathbf{A}$ of Theorem 11. So there is a subnet ν' of

ν which is eventually in every T-neighbourhood of a. Thus ν' converges to a.

(2) Conversely, suppose a is not an adherent point of ν.

Then there is a T-neighbourhood V of a such that ν is not frequently in V, and hence is eventually in $C_E(V)$. So every subnet ν' of ν is eventually in $C_E(V)$ and hence does not converge to a.

123. (1) Suppose F converges to x, so that $V(x) \subseteq F$.

Let ν be any net associated with F. Let V be any T-neighbourhood of x. For every set X in F which is R-greater than V we have $\nu(X) \in X \subseteq V$. So ν is eventually in V. Thus ν converges to x.

(2) Conversely, suppose F does not converge to x.

Then there is a T-neighbourhood V of x which does not belong to F. So for every set X in F we have $X \cap C_E(V) \neq \emptyset$; let a_X be any element of $X \cap C_E(V)$. Let ν be the net with domain F given by $\nu(X) = a_X$ for all X in F. Then ν does not converge to x, but is associated with F.

124. (1) Suppose ν converges to x.

Then ν is eventually in every T-neighbourhood of x. Thus every T-neighbourhood of x belongs to $F(\nu)$, i.e. $F(\nu) \supseteq V_T(x)$. So $F(\nu)$ converges to x.

(2) The converse is equally clear.

125. (1) Suppose f is (T, T')-continuous at x.

Let ν be a net in E which converges to x. Let V' be any T'-neighbourhood of $f(x)$. Then there is a T-neighbourhood V of x such that $f^{\rightarrow}(V) \subseteq V'$. Since ν converges to x, ν is eventually in V. So $f \circ \nu$ is eventually in $f^{\rightarrow}(V)$ and so in V'. So $f \circ \nu$ converges to $f(x)$.

(2) Conversely, suppose f is not (T, T')-continuous at x.

Then there is a T'-neighbourhood V' of $f(x)$ such that for every T-neighbourhood V of x the set $f^{\rightarrow}(V)$ is not included in V'. For each T-neighbourhood V of x let a_V be an element of V such that $f(a_V) \notin V'$. Consider the net ν with domain $V_T(x)$ given by $\nu(V) = a_V$ for each V in $V_T(x)$. Then ν converges to x but $f \circ \nu$ does not converge to $f(x)$.

126. (1) Suppose ν converges to a.

Let V be any T-neighbourhood of a. Since ν is eventually in V there is an element d of D such that $\nu(n) \in V$ for all $n \geq d$. Since $\nu' = \nu \circ \varphi$ is a subnet of ν there is an element d' of D' (where D' is the domain of ν') such that for all $n' \geq d'$ we have $\varphi(n') \geq d$ and so $\nu(\varphi(n')) = \nu'(n') \in V$. So ν' is eventually in V.

Thus ν' converges to a.

(2) Suppose a is adherent to ν'.

Let V be any T-neighbourhood of a. Let d be any element of D. Then there is an element d' of D' such that for all $n' \geq d'$ we have $\varphi(n') \geq d$. Since ν' is frequently in V there is an element n' in D' such that $n' \geq d'$ and $\nu'(n') \in V$. Then we have $\varphi(n') \geq d$ and $\nu(\varphi(n')) = \nu'(n') \in V$. So ν is frequently in V.

Thus a is adherent to ν.

127. (1) Let $\nu : D \to E$ be a net, $\nu' = \nu \circ \varphi : D' \to E$ a subnet.

Suppose ν is an ultranet. Let X be any subset of E. Then either (a) ν is eventually in X or (b) ν is eventually in $C_E(X)$. By the argument of Exercise 126 (1) we deduce that in case (a) ν' is eventually in X and in case (b) ν' is eventually in $C_E(X)$. Thus ν' is an ultranet.

(2) Let **S** be the collection of all sets Q of subsets of E such that (1) ν is frequently in every subset of E belonging to Q and (2) Q is closed under finite intersection.

Then **S** is inductively ordered by inclusion and so, by Zorn's Lemma, has a ($\subseteq$)-maximal element, Q_0 say. By Theorem 11, ν has a subnet ν' which is eventually in every subset of E belonging to Q_0. Let A be any subset of E. We claim that either A or its complement belongs to Q_0.

To see this, suppose first that for every set X in Q_0 the net ν is frequently in $A \cap X$. Then $Q_0 \cup \{A\} \supseteq Q_0$ and $Q_0 \cup \{A\}$ belongs to **S**. Hence, since Q_0 is ($\subseteq$)-maximal in **S**, we have $A \in Q_0$.

Suppose, on the other hand, that there is a set X_1 in Q_0 such that ν is eventually in $C_E(A \cap X_1)$. Then $Q_0 \cup \{C_E(A \cap X_1)\} \supseteq Q_0$ and belongs to **S**. Hence, since Q_0 is ($\subseteq$)-maximal in **S**, we have $C_E(A \cap X_1) \in Q_0$. Since $C_E(A) \supseteq C_{X_1}(A \cap X_1)$, it follows that $Q_0 \cup C_E(A) \in$ **S** and so, as above, $C_E(A) \in Q_0$.

Thus ν' is an ultranet.

128. (1) Let ν be an ultranet.
The filter associated with ν is

$$F(\nu) = \{X \in \mathbf{P}(E) : \nu \text{ is eventually in } X\}.$$

For every subset X of E we have either (1) ν is eventually in X or else (2) ν is eventually in $C_E(X)$, i.e. either $X \in F(\nu)$ or $C_E(X) \in F(\nu)$. Thus $F(\nu)$ is an ultrafilter.

(2) Suppose F is an ultrafilter.

Let ν be any net associated with F.

Let Y be any subset of E. Suppose ν is not eventually in Y. Then ν is frequently in $C_E(Y)$, i.e. for every set X in F there is a set X' in F such that $X' \subseteq X$ and $\nu(X') \in C_E(Y)$. Since $\nu(X') \in X'$ this shows that $X' \cap C_E(Y) \neq \emptyset$ and so $X \cap C_E(Y) \neq \emptyset$. We deduce that $F \cup \{C_E(Y)\}$ generates a filter F' which includes F. Since F is an ultrafilter we must have $F' = F$ and so $C_E(Y) \in F$. Then for every set X' in F such that $X' \subseteq C_E(Y)$ we have $\nu(X') \in X' \subseteq C_E(Y)$. So ν is eventually in $C_E(Y)$. Thus ν is an ultranet.

Chapter 12

ANSWERS FOR CHAPTER 5

129. Let T be the particular point topology on E determined by the point p.

Let a and b be distinct points of E. If one of these points, say a, is the particular point p, then $\{a\} = \{p\}$ is a T-neighbourhood of a which does not contain b. If a and b are both distinct from p then $\{a, p\}$ is a T-neighbourhood of a which does not contain b. So (E, T) is T_0.

130. (1) Suppose that T is T_0.

Let x and y be distinct points of E. Since T is T_0 there is either a T-neighbourhood V_x of x which does not contain y or a T-neighbourhood V_y of y which does not contain x. In the first case $x \in \mathrm{Cl}_T\{x\}$ but $x \notin \mathrm{Cl}_T\{y\}$; in the second case $y \in \mathrm{Cl}_T\{y\}$ but $y \notin \mathrm{Cl}_T\{x\}$. So in each case we have $\mathrm{Cl}_T\{x\} \neq \mathrm{Cl}_T\{y\}$.

(2) Conversely, suppose that for each pair of distinct points x, y of E we have $\mathrm{Cl}_T\{x\} \neq \mathrm{Cl}_T\{y\}$.

Suppose that (E, T) is not a T_0 space. Then there is a pair of distinct points a, b of E such that every T-neighbourhood of a contains b and every T-neighbourhood of b contains a. Thus $a \in \mathrm{Cl}\{b\}$ and $b \in \mathrm{Cl}\{a\}$. So $\{a\} \subseteq \mathrm{Cl}\{b\}$ and $\{b\} \subseteq \mathrm{Cl}\{a\}$; it follows that $\mathrm{Cl}\{b\} \subseteq \mathrm{Cl}\{a\}$ and $\mathrm{Cl}\{a\} \subseteq \mathrm{Cl}\{b\}$. Hence $\mathrm{Cl}\{a\} = \mathrm{Cl}\{b\}$, which is a contradiction. So (E, T) is T_0.

131. Let η be the canonical surjection from E onto E/R. We claim that η is $(T, T/R)$-open.

So let U be any T-open subset of E. Then we have $\eta^{\leftarrow}(\eta^{\rightarrow}(U)) = \bigcup_{x \in U} R(x)$, where $R(x)$ is the R-class of x. Suppose x is any point of U and $t \in R(x)$. Then $\mathrm{Cl}\{t\} = \mathrm{Cl}\{x\}$ and so $x \in \mathrm{Cl}\{t\}$; hence every T-open set which contains x also contains t. In particular, $t \in U$. Hence $U \subseteq \eta^{\leftarrow}(\eta^{\rightarrow}(U)) \subseteq U$. So $\eta^{\leftarrow}(\eta^{\rightarrow}(U))$ is T-open and hence $\eta^{\rightarrow}(U)$ is T/R-open. Thus η is $(T, T/R)$-open.

Let now X and Y be distinct points of E/R, say $X = \eta(x)$ and $Y = \eta(y)$, with $x \in E$, $y \in E$.

Then $(x, y) \notin R$ and hence $\mathrm{Cl}_T\{x\} \neq \mathrm{Cl}_T\{y\}$. Hence there is either an open set containing x but not y or an open set containing y but not x; say there is an open set U such that $x \in U$, $y \notin U$.

Then $\eta^{\rightarrow}(U)$ is a (T/R)-open set containing $X = \eta(x)$. We shall show that $\eta^{\rightarrow}(U)$ does not contain $Y = \eta(y)$. If $Y = \eta(y) \in \eta^{\rightarrow}(U)$, then there is a point t of U such that $\eta(y) = \eta(t)$, from which we have $\mathrm{Cl}_T\{y\} = \mathrm{Cl}_T\{t\}$ and so $t \in \mathrm{Cl}_T\{y\}$. Now U is a T-open set containing t; so $U \cap \{y\} \neq \emptyset$, i.e. $y \in U$, which contradicts the original choice of U. Thus $(E/R, T/R)$ is T_0.

132. (1) Suppose p is a metric.

Let x and y be distinct points of E. Let $r = p(x, y)$; since p is a metric, r is non-zero. Then $V_p(x, r)$ is a T_p-neighbourhood of x which does not contain y.

(2) Conversely, if p is not a metric there are distinct points x, y such that $p(x, y) = 0$. Then for every positive real number r we have $y \in V_p(x, r)$ and $x \in V_p(y, r)$. It follows that the topology T_p is not T_0.

133. (1) Let x be any point of E. Since $x \in \mathrm{Cl}_T\{x\}$ we have $(x, x) \in A$. Thus A is reflexive.

(2) Let x, y, z be points of E such that $(x, y) \in A$ and $(y, z) \in A$. To show that $(x, z) \in A$, i.e. that $x \in \mathrm{Cl}_T\{z\}$ let V be any T-neighbourhood of x. Then V includes a T-open subset U containing x. Since $x \in \mathrm{Cl}_T\{y\}$ the set U meets $\{y\}$, i.e. $y \in U$, so U is a T-neighbourhood of y. Since $y \in \mathrm{Cl}_T\{z\}$ it follows that U meets $\{z\}$. So V meets $\{z\}$ and hence $x \in \mathrm{Cl}_T\{z\}$ as required. Thus A is transitive.

(3) Suppose (E, T) is a T_0-space.

If $(x, y) \in A$ and $(y, x) \in A$ then we have $x \in \mathrm{Cl}_T\{y\}$ and also

$y \in \mathrm{Cl}_T\{x\}$. It follows that $\mathrm{Cl}_T\{x\} = \mathrm{Cl}_T\{y\}$ and so, by Exercise 130, we have $x = y$. Thus A is antisymmetric.

Conversely, suppose A is antisymmetric.

Let x and y be distinct points of E. Then we cannot have both $(x,y) \in A$ and $(y,x) \in A$; so we cannot have both $y \in \mathrm{Cl}_T\{x\}$ and $x \in \mathrm{Cl}_T\{y\}$. Thus $\mathrm{Cl}_T\{x\} \neq \mathrm{Cl}_T\{y\}$. Hence (E,T) is T_0 by Exercise 130.

134. (1) Suppose K is T-closed.

If $y \in K$ and $(x,y) \in A$ then $x \in \mathrm{Cl}_T\{y\}$. But since $\{y\} \subseteq K$ we have $\mathrm{Cl}_T\{y\} \subseteq \mathrm{Cl}_T K = K$. So $x \in K$.

(2) Conversely, let K be a subset of E such that whenever $y \in K$ and $(x,y) \in A$ we have $x \in K$.

Since (E,T) is an Alexandrov space the union of every family of T-closed subsets of E is T-closed. It follows readily that $\mathrm{Cl}_T K = \bigcup_{y \in K} \mathrm{Cl}_T\{y\}$. Thus if x is any point of $\mathrm{Cl}_T K$ we must have $x \in \mathrm{Cl}_T\{y\}$, i.e. $(x,y) \in A$, for some point y of K. But this implies that $x \in K$. So $\mathrm{Cl}_T K = K$ and hence K is T-closed, as required.

135. We have seen in Exercise 129 that each particular point topology is T_0. If T is the particular point topology determined by the point p and a is any point distinct from p, every T-open set containing a contains p also. Thus T is not T_1.

136. If k is odd, say $k = 2n + 1$, then $\{k\}$ is the intersection of the T-open sets $\{2n - 1, 2n, 2n + 1\}$ and $\{2n + 1, 2n + 2, 2n + 3\}$ and hence is T-open.

If k is even, say $k = 2n$, then $\{k\}$ is the complement of the union of the family of all the T-open sets $\{2m - 1, 2m, 2m + 1\}$ with $m \neq n$. This union is T-open; hence $\{k\}$ is T-closed.

To show that the digital topology is T_0 let x and y be distinct integers. If one of them is odd, say x, then $\{x\}$ is a T-neighbourhood of x which does not contain y. If both x and y are even then $\{x-1, x, x+1\}$ is a T-neighbourhood of x which does not contain y.

The digital topology is not T_1 since there is no T-neighbourhood of $2n$ which fails to contain $2n + 1$.

137. Let x and y be distinct points of E.

Let $r = q(x,y)$ and $s = q(y,x)$. Then $V_q(x,r)$ is a T_q-neighbourhood of x which does not contain y and $V_q(y,s)$ is a T_q-neighbourhood of y which does not contain x.

138. (1) $\Longrightarrow$ (2) Suppose T is a T_1 topology

Let x be any point of E. We claim that $\mathrm{Cl}\{x\} = \{x\}$.

If t is any point of E distinct from x there is a T-open set U containing t which does not contain x. So t is not an adherent point of $\{x\}$. Thus $t \notin \mathrm{Cl}\{x\}$. It follows that $\mathrm{Cl}\{x\} = \{x\}$ and so $\{x\}$ is T-closed.

(2) $\Longrightarrow$ (3) Suppose that for every point t of E the set $\{t\}$ is T-closed.

Let x be any point of E, y any point of $\bigcap N(x)$, where $N(x)$ is the T-neighbourhood filter of x. Then every T-neighbourhood of x contains y and so meets $\{y\}$. Thus $x \in \mathrm{Cl}_T\{y\} = \{y\}$. So $y = x$. Hence $\bigcap N(x) = \{x\}$.

(3) $\Longrightarrow$ (1) Suppose that for every point t of E we have $\bigcap N(t) = \{t\}$.

Let x and y be distinct points of E. Since $y \notin \{x\} = \bigcap N(x)$ there is a T-neighbourhood of x which does not contain y. Similarly there is a T-neighbourhood of y which does not contain x. Thus T is T_1.

139. Let V be any T-neighbourhood of x.

Suppose V contains only finitely many points of A, say $a_1, \ldots, a_n$. Then there are neighbourhoods $V_1, \ldots, V_n$ of x such that $a_i \notin V_i$ $(i = 1, \ldots, n)$. Then $V \cap V_1 \cap V_2 \cap \ldots \cap V_n$ is a T-neighbourhood of x containing no point of A. This is a contradiction, since x is an adherent point of A. So V must contain infinitely many points of A.

140. Let x be a point of E.

There is a set B_1 in the finite base such that $x \in B_1$. If $B_1 \neq \{x\}$ there is a point $x_1 \neq x$ such that $x_1 \in B_1$. Since T is T_1 there is a T-neighbourhood V of x not containing x_1. Then $V \cap B_1$ must include a basic open set B_2 which contains x but not x_1. Thus $B_1 \supseteq B_2$.

If $B_2 \neq \{x\}$ we may repeat the process, so obtaining a strictly decreasing sequence of basic open sets. Since there are only finitely many of these basic open sets the process must stop, i.e. we must eventually have $\{x\}$ = one of the basic open sets. Thus the topology is discrete; and since there are only finitely many basic open sets there can be only finitely many points of E.

141. This is an immediate consequence of Exercise 140 since a finite set has only finitely many subsets; hence every topology on a finite set has a finite base.

142. Let E be an infinite set, T the finite complement topology on E.

If a and b are distinct points of E then $C_E\{a\}$ is a T-open set which contains b but not a and $C_E\{b\}$ is a T-open set which contains a but not b. Thus (E, T) is T_1.

Suppose (E, T) were Hausdorff.

Then if a and b are distinct points of E there would be disjoint T-open sets U and V containing a and b respectively. Then we would have $E = C_E(U \cap V) = C_E(U) \cup C_E(V)$ which is finite, and this is a contradiction. So (E, T) is not Hausdorff.

143. Let x and y be distinct points of E. Let $r = d(x, y)$. Then $V_d(x, \frac{1}{2}r)$ and $V_d(y, \frac{1}{2}r)$ are disjoint neighbourhoods of x and y respectively. For, if there were a point z in their intersection, we would have $d(x, z) < \frac{1}{2}r$, $d(y, z) < \frac{1}{2}r$ and hence $r = d(x, y) \leq d(x, z) + d(z, y) = d(x, z) + d(y, z) < r$ which is a contradiction.

144. The verification of the Triangle Inequality $q(x, z) \leq q(x, y) + q(y, z)$ involves a tedious, but entirely routine, examination of cases (depending on whether each of x, y, z is 0, ∞ or a positive integer).

To show that (E, T) is not Hausdorff, let U and V be T_q-neighbourhoods of 0 and ∞ respectively. Then there are positive real numbers r and s such that $V_q(0, r) \subseteq U$ and $V_q(\infty, s) \subseteq V$. Let n be any integer greater than the larger of $1/r$ and $1/s$. Then $q(0, n) = q(\infty, n) = 1/n$ is less than both r and s. So U and V are not disjoint and so (E, T) is not Hausdorff.

145. Let (x_1, y_1), (x_2, y_2) be distinct points of E.

Since x_1, y_1, x_2, y_2 are rational and θ is irrational, $x_1 - \theta y_1$ and

$x_2 - \theta y_2$ are distinct and $x_1 + \theta y_1$ and $x_2 + \theta y_2$ are distinct. Let

$$\varepsilon = \min\{\,|\,(x_1 - \theta y_1) - (x_2 - \theta y_2)\,|,\ |\,(x_1 + \theta y_1) - (x_2 + \theta y_2)\,|\,\}.$$

Then $N_{\varepsilon/2}(x_1, y_1)$ and $N_{\varepsilon/2}(x_2, y_2)$ are disjoint T_θ-neighbourhoods of (x_1, y_1), (x_2, y_2) respectively. Thus (E, T_θ) is Hausdorff.

146. (1) $\implies$ (2) Suppose (E, T) is Hausdorff.

Let x be any point of E, y any point distinct from x. There exist disjoint T-neighbourhoods V of x and W of y. Since $W \cap V = \emptyset$ it follows that $y \notin \mathrm{Cl}\,V$, which is a closed T-neighbourhood of x. Thus y is not in every closed T-neighbourhood of x. So the intersection of the family of closed T-neighbourhoods of x is $\{x\}$.

(2) $\implies$ (3) Suppose condition (2) holds.

Let F be a filter on E which converges to x, i.e. $F \supseteq N(x)$, the T-neighbourhood filter of x. If y is adherent to F, then y belongs to the closure of every T-neighbourhood of x and hence to every closed T-neighbourhood of x.

Thus $y = x$, i.e. x is the only adherent point of F.

(3) $\implies$ (4) Suppose condition (3) holds.

Let F be a filter on E. If x and y are limit points of F they are also adherent points and so, by condition (3), $x = y$.

(4) $\implies$ (1) Suppose condition (4) holds.

If T is not Hausdorff there is a pair of distinct points x and y such that every T-neighbourhood of x meets every T-neighbourhood of y. Thus $N(x) \cup N(y)$ generates a filter F which converges to both x and y. This is a contradiction. So T is Hausdorff.

147. (Specimen only) Suppose (E, T) is T_2 (Hausdorff).

Let a_1 and a_2 be distinct points of a subset A of E. Then there are disjoint T-open subsets U_1, U_2 of E containing a_1, a_2 respectively. Then $A \cap U_1$ and $A \cap U_2$ are disjoint T_A-open subsets of A containing a_1 and a_2 respectively. So (A, T_A) is Hausdorff.

148. (Specimens only) Suppose all the spaces (E_i, T_i) are T_0.

Let x and y be distinct points of $E = \prod E_i$. Then there is an index j in I such that $\pi_j(x) \neq \pi_j(y)$. Hence there is a T_j-neighbourhood of one of these points which does not contain the other, say $\pi_j(x) \in V_j$, $\pi_j(y) \notin V_j$. For $i \neq j$ let $V_i = E_i$. Then $\prod_{i \in I} V_i$ is a T-neighbourhood of x which does not contain y. Thus (E, T) is T_0.

Suppose the product (E, T) is Hausdorff.

Let x_{i_0}, y_{i_0} be distinct points of E_{i_0}. Let x and y be points of E such that $\pi_{i_0}(x) = x_{i_0}$, $\pi_{i_0}(y) = y_{i_0}$ and $\pi_j(x) = \pi_j(y)$ for all $j \neq i_0$. Then there are disjoint T-neighbourhoods U and V of x and y respectively. There are finite subsets J and K of I and families $(U_i)_{i \in I}$, $(V_i)_{i \in I}$ of T_i-open subsets of E_i such that $U_i = E_i$ for $i \notin J$ and $V_i = E_i$ for $i \notin K$, and $x \in \prod_{i \in I} U_i \subseteq U$, $y \in \prod_{i \in I} V_i \subseteq V$. Then U_{i_0}, V_{i_0} are disjoint T_{i_0}-neighbourhoods of x_{i_0}, y_{i_0} respectively. Thus (E_{i_0}, T_{i_0}) is Hausdorff.

149. Let $P = C_{E \times E}(D)$.

(1) Suppose (E, T) is Hausdorff.

Let (x, y) be any point of P. Since $x \neq y$ there are disjoint T-open subsets U and V of E containing x and y respectively. Then $U \times V$ is a $T \times T$-neighbourhood of (x, y) which does not meet D. (For if $(a, a) \in D \cap (U \times V)$ we have $a \in U \cap V = \emptyset$, a contradiction.) So $(x, y) \in U \times V \subseteq P$. So P is a $(T \times T)$-neighbourhood of (x, y). Thus P is $(T \times T)$-open and so D is $(T \times T)$-closed.

(2) Conversely, suppose D is $(T \times T)$-closed.

Let x and y be distinct points of E. Then (x, y) belongs to the $(T \times T)$-open set P. Hence there are T-open sets U and V of E such that $(x, y) \in U \times V \subseteq P$. Then U and V are disjoint T-open sets containing x and y respectively. So (E, T) is Hausdorff.

150. Let $A = \{x \in E : f(x) = g(x)\}$.

Let t be any point not in A; then $f(t) \neq g(t)$. Since (E', T') is Hausdorff there are disjoint T'-open sets U', V' containing $f(t)$, $g(t)$ respectively. Since f and g are continuous there are T-open sets U, V containing t such that $f^{\rightarrow}(U) \subseteq U'$ and $g^{\rightarrow}(V) \subseteq V'$. Then $U \cap V \subseteq C_E(A)$. For if there were a point a of A in $U \cap V$ we would have $f(a) \in f^{\rightarrow}(U) \subseteq U'$, $g(a) \in g^{\rightarrow}(V) \subseteq V'$ and so $f(a) = g(a) \in U' \cap V' = \emptyset$. Thus $C_E(A)$ is a T-neighbourhood of each of its points t. So $C_E(A)$ is T-open and A is T-closed.

151. Let G be the graph of f, G' the complement of G in $E \times E'$.

Let (x, y) be any point of G'. Then $y \neq f(x)$. Since (E', T') is Hausdorff there are disjoint T'-open subsets U' and V' of E' such that $f(x) \in U'$ and $y \in V'$. Since f is (T, T')-continuous there is a T-open subset U of E containing x such that $f^{\rightarrow}(U) \subseteq U'$. Then $U \times V'$ is a $(T \times T')$-neighbourhood of (x,y). Let (a,b) be any point of $U \times V'$. Then $f(a) \in f^{\rightarrow}(U) \subseteq U'$ and $b \in V'$; so $f(a) \neq b$. Hence we have $(x,y) \in U \times V' \subseteq G'$. Thus G' is $(T \times T')$-open and so G is $(T \times T')$-closed.

152. Let $E = \{x_1, x_2, \ldots, x_n\}$.

Since (E, T) is Hausdorff there are T-open sets $U_2, \ldots, U_n$ containing x_1 but not $x_2, \ldots, x_n$ respectively. Then $U_2 \cap \ldots \cap U_n$ is a T-open set containing x_1 but not $x_2, \ldots, x_n$. Thus $\{x_1\}$ is T-open. Similarly $\{x_2\}, \ldots, \{x_n\}$ are T-open. Thus T is discrete.

153. We showed in Exercise 145 that the irrational slope topology T_θ is Hausdorff.

Let (x,y) be any point of E, ε any positive real number. Then
$$\mathrm{Cl}_{T_\theta}(N_\varepsilon(x,y)) \supseteq$$
$$\{(r,s) \in E : |(r - \theta s) - (x - \theta y)| < \varepsilon\}$$
$$\cup \{(r,s) \in E : |(r + \theta s) + (x + \theta y)| < \varepsilon\}.$$
It follows that if (x,y) and (x',y') are distinct points of E and V, V' are any T_θ-neighbourhoods of (x,y), (x',y') respectively then we have $\mathrm{Cl}\, V \cap \mathrm{Cl}\, V' \neq \emptyset$. So (E, T_θ) is not completely Hausdorff.

154. Let p and q be distinct points of E.

Let d be the ordinary Euclidean distance between p and q. Then $\mathrm{Cl}\,(D(p, \tfrac{1}{3}d)) \cap \mathrm{Cl}\,(D(q, \tfrac{1}{3}d)) = \emptyset$.

155. Let $p = (a, 0)$ be a point of the horizontal axis. Let F be the complement with respect to E of the set
$$\{p\} \cup \{(x,y) \in E : y > 0 \text{ and } (x - a)^2 + y^2 < 1\}.$$

Then F is a T-closed set and $p \notin F$. F contains all points of the form $(t, 0)$ where $a - 1 < t < a + 1$. Any open set containing p contains some such points, hence cannot be disjoint from an open set which includes F. Thus T is not regular.

156. Let $k = 2n$ be an even integer. Then $\{k\}$ is a closed set in the digital topology. Of course $2n + 1$ does not belong to this closed set. But the smallest open set including $\{2n\}$ is $\{2n - 1, 2n, 2n + 1\}$. So we cannot find an open set including $\{2n\}$ which fails to include $2n + 1$.

So the digital topology is not regular.

157. The only T-closed subsets are $\emptyset$, E, $\{a\}$, $\{b, c\}$. Thus the only ordered pairs (x, F) consisting of a point x and a T-closed set F not containing x are $(a, \emptyset)$, $(b, \emptyset)$, $(c, \emptyset)$, $(a, \{b, c\})$, $(b, \{a\})$, $(c, \{a\})$. In each case it is clear how to find disjoint open set U, V such that $x \in U$, $F \subseteq V$. So T is regular.

T is not T_1 since $\{b\}$ and $\{c\}$ are not T-closed.

158. (1) Suppose T is regular.

Let x be a point of E, U a T-open set containing x. Then $C_E(U)$ is a T-closed set not containing x. Hence there are disjoint T-open sets U' and V' such that $x \in U'$ and $C_E(U) \subseteq V'$. Since $U' \cap V' = \emptyset$ we have $U' \subseteq C_E(V') \subseteq U$.

Since $C_E(V')$ is T-closed we have $\operatorname{Cl} U' \subseteq C_E(V') \subseteq U$.

(2) Conversely, suppose the condition holds.

Let x be a point of E, F a T-closed set not containing x. Then $C_E(F)$ is a T-open set containing x. By hypothesis there is a T-open set U' such that $x \in U'$ and $\operatorname{Cl} U' \subseteq C_E(F)$. Let V' be the complement of $\operatorname{Cl} U'$; V' is T-open and we clearly have $x \in U'$, $F \subseteq V'$ and $U' \cap V' = \emptyset$. So T is regular.

159. (1) Suppose T is regular.

Let x be any point of E, V any T-neighbourhood of x. Then there exists a T-open set U such that $x \in U \subseteq V$. Since T is regular it follows from Exercise 158 that there is a T-open set U' such that $x \in U' \subseteq \operatorname{Cl} U' \subseteq U \subseteq V$. The collection of all closed sets $\operatorname{Cl} U'$ obtained in this way is a fundamental system of T-neighbourhoods.

(2) Conversely, suppose every point of E has a T-neighbourhood base consisting of T-closed sets.

Let x be any point of E, U a T-open set containing x. Then U is a T-neighbourhood of x. By hypothesis there is a closed T-neighbourhood F of x such that $x \in F \subseteq U$. Since F is a T-neighbourhood of x there exists a T-open set U' such that $x \in U' \subseteq F$. Then we have $x \in U'$ and $\operatorname{Cl} U' \subseteq F \subseteq U$. So T is regular.

160. Suppose (E, T) is a regular space, A a subset of E.

Let a be a point of A, F a T_A-closed subset of A not containing a. Then there is a T-closed subset F' of E such that $F = F' \cap A$. Since $a \notin F'$ and (E, T) is regular there are disjoint T-open sets U, V such that $a \in U$ and $F \subseteq V$. Then $A \cap U$ and $A \cap V$ are disjoint T_A-open sets with $a \in A \cap U$ and $F \subseteq A \cap V$. Thus (A, T_A) is regular.

161. (1) Let p and q be distinct points of E.

If p and q are not on the horizontal axis let d be the ordinary Euclidean distance from p to q. Then $E \cap V(p, d)$ is a T-open set containing p but not q and $E \cap V(q, d)$ is a T-open set containing q but not p.

If $p = (x, 0)$ and q are on the horizontal axis then for each positive number d the set $\{p\} \cup V((x, d), d)$ is a T-open set containing p but not q. Similarly there are T-open sets containing q but not p.

If $p = (x, 0)$ is on the horizontal axis and $q = (s, t)$ is not then $\{p\} \cup V((x, \frac{1}{2}t), \frac{1}{2}t)$ is a T-open set containing p which does not contain q and $V(q, t)$ is a T-open set containing q which does not contain p. Thus T is T_1.

(2) Let F be a T-closed subset of E, p a point not in F.

If $p = (x, y)$ is not on the horizontal axis there is a positive real number r such that $V(p, r) \subseteq C_E(F)$. Define $f : E \to [0, 1]$ by setting $f(t) = 1$ for all $t \notin V(p, r)$ and $f(t) = d(t, p)/r$ for all points $t \in V(p, r)$. Then f is continuous, $f(p) = 0$ and $f(t) = 1$ for all t in F. ($d(t, p)$ is the ordinary Euclidean distance from t to p.)

If $p = (x, 0)$ is on the horizontal axis there is a positive real number r such that $\{p\} \cup V((x, r), r) \subseteq C_E(F)$. Define $f : E \to [0, 1]$ by setting $f(p) = 0$, $f(t) = 1$ for all $t \notin \{p\} \cup V((x, r), r)$ and $f(t) = d(t, p)/d(q_t, p)$ for all points $t \in V((x, r), r)$. (Here q_t is the point where the line joining p and t meets the circumference of $V((x, r), r)$ and d is the ordinary Euclidean metric.) Then f is continuous, $f(p) = 0$ and $f(t) = 1$ for all t in F. So (E, T) is completely regular and hence $T_{3\frac{1}{2}}$.

162. Let (E, T) be a completely regular space.

To show that it is regular let a be a point of E, F a closed subset of E not containing a. Since (E, T) is completely regular there exists a continuous mapping f from E to $[0, 1]$ such that $f(a) = 0$ and $f(t) = 1$ for all points t of F. $[0, \frac{1}{4})$ and $(\frac{3}{4}, 1]$ are open subsets of $[0, 1]$. Hence

$f^{\leftarrow}[0, \frac{1}{4})$ and $f^{\leftarrow}(\frac{3}{4}, 1]$ are open subsets of E which are clearly disjoint. We also have $a \in f^{\leftarrow}[0, \frac{1}{4})$ and $F \subseteq f^{\leftarrow}(\frac{3}{4}, 1]$. So (E, T) is regular.

163. Let $((E_i, T_i))_{i \in I}$ be a family of completely regular spaces, (E, T) its product.

Let a be a point of E, F a closed subset of E not containing a. Then $C_E(F)$ is T-open and hence there is a family $(U_i)_{i \in I}$ of T_i-open sets and a finite subset J of I such that $U_i = E_i$ for all $i \notin J$ and $a \in \prod_{i \in I} U_i \subseteq C_E(F)$. For each index i in J, there exists a continuous mapping f_i from E_i to $[0, 1]$ such that $f_i(\pi_i(a)) = 0$ and $f_i(t_i) = 1$ for all $t_i \in C_{E_i}(U_i)$. Let $f = \sup_{i \in J}(f_i \circ \pi_i)$. Then f is a continuous mapping from E to $[0, 1]$, $f(a) = 0$ and $f(t) = 1$ for all t in F. Thus (E, T) is completely regular.

164. (1) Since $[0, 1]$ is easily seen to be $T_{3\frac{1}{2}}$ so is any product of spaces homeomorphic to $[0, 1]$ (by Exercise 163) and it is easy to show that every subspace of a $T_{3\frac{1}{2}}$ space is $T_{3\frac{1}{2}}$.

(2) Let (E, T) be a $T_{3\frac{1}{2}}$ space.

Let F be the set of continuous mappings from E to $[0, 1]$. For each f in F let $I_f = [0, 1]$; let $P = \prod_{f \in F} I_f$. Define the mapping g from E to P by setting $g(x) = (f(x))_{f \in F}$ for all x in E.

(1) We claim that g is injective.

So let x and y be distinct points of E. Since (E, T) is T_1, $\{y\}$ is a T-closed set which does not contain x. So, since T is $T_{3\frac{1}{2}}$, there is a continuous mapping f_0 from E to $[0, 1]$ such that $f_0(x) = 0$ and $f_0(y) = 1$. Hence $g(x)$ and $g(y)$ differ in the f_0-th coordinate and so $g(x) \neq g(y)$.

(2) Next we show that g is continuous.

Let x be any point of E, V' any neighbourhood of $g(x)$ in P.

Then there is a finite subset G of F and a family $(V'_f)_{f \in F}$ of open subsets of $[0, 1]$ such that we have $V'_f = [0, 1]$ for all $f \notin G$ and further $g(x) \in \prod_{f \in F} V'_f \subseteq V'$. For each mapping f in G, f is continuous and so there is a T-neighbourhood V_f of x in E such that $f^{\rightarrow}(V_f) \subseteq V'_f$. Let $V = \cap_{f \in G} V_f$; V is a T-neighbourhood of x and $g^{\rightarrow}(V) \subseteq V'$.

(3) Finally we show that the restriction of g^{-1} to $g^{\rightarrow}(E)$ is continuous.

Let $p = g(x)$ be any point of $g^{\rightarrow}(E)$, V any T-neighbourhood of

$g^{-1}(p) = x$. Then V includes a T-open subset U of E containing x. Since (E, T) is completely regular there is a continuous mapping f_1 from E to $[0, 1]$ such that $f_1(x) = 0$ and $f_1(y) = 1$ for all points y of $C_E(U)$ and hence for all points y of $C_E(V)$. Let $V'_{f_1} = [0, 1)$, $V'_f = [0, 1]$ for all $f \neq f_1$. Then $V' = \prod_{f \in F} V'_f$ is a neighbourhood of $p = g(x)$ in P.

We claim that $(g^{-1})^{\rightarrow}(V' \cap g^{\rightarrow}(E)) \subseteq V$. So let t be any point of $(g^{-1})^{\rightarrow}(V' \cap g^{\rightarrow}(E))$. Then $g(t) \in V'$. If $t \notin V$ then $f_1(t) = 1$ and so $g(t) \notin V'$. Hence $t \in V$.

Thus g is a homeomorphism from E onto $g^{\rightarrow}(E) \subseteq \prod_{f \in F} I_f$.

165. Let $2n$ and $2n + 2$ be successive even integers. Then $\{2n\}$ and $\{2n + 2\}$ are closed in the digital topology. But the smallest open sets of the digital topology including them are $\{2n - 1, 2n, 2n + 1\}$ and $\{2n + 1, 2n + 2, 2n + 3\}$ respectively. It follows that we cannot find disjoint open sets including the disjoint closed sets $\{2n\}$, $\{2n + 2\}$. So the digital topology is not normal.

166. Compare Exercise 140.

167. Let $U_1 = C_E(B)$. Then U_1 is a T-open subset of E which includes A. According to Exercise 158 there is a T-open subset U_0 such that $A \subseteq U_0$ and $\mathrm{Cl}\, U_0 \subseteq U_1$.

Suppose now that for some natural number k we have constructed a sequence of T-open sets $(U_{i/2^k})_{0 \leq i \leq 2^k}$ such that for $i = 0, 1, \ldots,$ $2^k - 1$ we have $\mathrm{Cl}\, U_{i/2^k} \subseteq U_{(i+1)/2^k}$. (We have certainly achieved this in the case $k = 0$.) It follows, again from Exercise 158, that for $j = 0, 1, \ldots, 2^k - 1$ there exists a T-open set $U_{(2j+1)/2^{k+1}}$ such that

$$\mathrm{Cl}\, U_{j/2^k} = \mathrm{Cl}\, U_{2j/2^{k+1}} \subseteq U_{(2j+1)/2^{k+1}}$$

and

$$\mathrm{Cl}\, U_{(2j+1)/2^{k+1}} \subseteq U_{(j+1)/2^k} = U_{(2j+2)/2^{k+1}}.$$

By induction it follows that for each dyadic rational r (rational number with denominator a power of 2) in $[0, 1]$ there is an open set U_r such that whenever r_1 and r_2 are dyadic rationals with $r_1 < r_2$ we have $\mathrm{Cl}\, U_{r_1} \subseteq U_{r_2}$.

Now define $f : E \to [0,1]$ by setting

$$f(x) = \begin{cases} 1 & \text{if } x \in B \\ \inf \{r \in [0,1] \,:\, x \in U_r\} & \text{if } x \in U_1 = C_E(B). \end{cases}$$

We shall show that f is continuous.

Since the topology of $[0,1]$ is generated by all intervals of the forms $[0,s)$ and $(t,1]$ it will be enough to show that the inverse images under f of such intervals are T-open.

(1) We claim that $f^{\leftarrow}[0,s) = \bigcup_{r \in K} U_r$ where K is the set of dyadic rationals r such that $0 \leq r < s$.

So suppose $x \in f^{\leftarrow}[0,s)$. Then $f(x) < s$ and there is a dyadic rational r_1 such that $f(x) \leq r_1 < s$. Thus $x \in U_{r_1}$. Hence we have $f^{\leftarrow}[0,s) \subseteq \bigcup_{r \in K} U_r$.

Conversely, suppose $x \in \bigcup_{r \in K} U_r$. Then $x \in U_{r_1}$ say, where r_1 is a dyadic rational such that $0 \leq r_1 < s$. Then $f(x) \leq r_1 < s$ and so $x \in f^{\leftarrow}[0,s)$. So $\bigcup_{r \in K} U_r \subseteq f^{\leftarrow}[0,s)$.

Hence $f^{\leftarrow}[0,s) = \bigcup_{r \in K} U_r$ and hence is open.

(2) Next we claim that $f^{\leftarrow}(t,1] = \bigcup_{r \in L} (C_E(\mathrm{Cl}\, U_r))$ where L is the set of dyadic rationals r such that $t < r \leq 1$.

So suppose $x \in f^{\leftarrow}(t,1]$. Then $f(x) > t$, and there are dyadic rationals r_1, r_2 such that $t < r_1 < r_2 < f(x)$. Then $x \notin U_{r_2}$ and so, since $\mathrm{Cl}\, U_{r_1} \subseteq U_{r_2}$ it follows that $x \notin \mathrm{Cl}\, U_{r_1}$. So $x \in C_E(\mathrm{Cl}\, U_{r_1})$. Thus $f^{\leftarrow}(t,1] \subseteq \bigcup_{r \in L} (C_E(\mathrm{Cl}\, U_r))$.

Conversely, suppose $x \in \bigcup_{r \in L} (C_E(\mathrm{Cl}\, U_r))$, say $x \notin U_{r_1}$ where $r_1 \in L$. Thus $f(x) \geq r_1 > t$ and so $x \in f^{\leftarrow}(t,1]$. Hence we have $\bigcup_{r \in L} (C_E(\mathrm{Cl}\, U_r)) \subseteq f^{\leftarrow}(t,1]$.

Hence $f^{\leftarrow}(t,1] = \bigcup_{r \in L} (C_E(\mathrm{Cl}\, U_r))$, which is T-open.

It follows that f is continuous and clearly we have $f^{\to}(A) = \{0\}$ and $f^{\to}(B) = \{1\}$.

168. Take $I = [-1,1]$. Let

$$\begin{aligned} A &= f^{\leftarrow}[-1,-\tfrac{1}{3}] = \{x \in F : -1 \leq f(x) \leq -\tfrac{1}{3}\}, \\ B &= f^{\leftarrow}[\tfrac{1}{3},1] = \{x \in F : \tfrac{1}{3} \leq f(x) \leq 1\}. \end{aligned}$$

Since f is continuous and $[-1,-\tfrac{1}{3}]$ and $[\tfrac{1}{3},1]$ are closed, A and B are T_F-closed and hence T-closed; they are clearly disjoint.

By an obvious modification of Urysohn's Lemma there exists a continuous mapping g from E to $[-\tfrac{1}{3},\tfrac{1}{3}]$ such that $g^{\to}(A) = \{-\tfrac{1}{3}\}$,

$g^{\rightarrow}(B) = \{\frac{1}{3}\}$. Set $f_0 = f$, $g_0 = g$, $f_1 = (f_0 - g_0)|F$. Then we claim that $|f_1(x)| \leq \frac{2}{3}$ for all x in F. This follows from consideration of the following three cases:

(a) If $x \in A$ then $-1 \leq f(x) = f_0(x) \leq -\frac{1}{3}$ and $g(x) = -\frac{1}{3}$; so $-\frac{2}{3} \leq f_0(x) - g_0(x) \leq 0$.

(b) If $x \in B$ then $\frac{1}{3} \leq f(x) = f_0(x) \leq 1$ and $g(x) = \frac{1}{3}$; so $0 \leq f_0(x) - g_0(x) \leq \frac{2}{3}$.

(c) If $x \in C_F(A \cup B)$ then $-\frac{1}{3} \leq f(x) \leq \frac{1}{3}$ and $-\frac{1}{3} \leq g(x) \leq \frac{1}{3}$; so we have $-\frac{2}{3} \leq f_0(x) - g_0(x) \leq \frac{2}{3}$.

Now let $A_1 = f_1^{\leftarrow}[-\frac{2}{3}, -\frac{1}{3} \cdot \frac{2}{3}]$ and $B_1 = f_1^{\leftarrow}[\frac{1}{3} \cdot \frac{2}{3}, \frac{2}{3}]$. By the same argument there exists a continuous mapping g_1 from E to $[-\frac{1}{3} \cdot \frac{2}{3}, \frac{1}{3} \cdot \frac{2}{3}]$ such that $g_1^{\rightarrow}(A_1) = \{-\frac{1}{3} \cdot \frac{2}{3}\}$ and $g_1^{\rightarrow}(B_1) = \{\frac{1}{3} \cdot \frac{2}{3}\}$, and if we set $f_2 = (f_1 - g_1)|F$ we have $|f_2(x)| \leq (\frac{2}{3})^2$ for all points x of F.

Proceeding in this way we obtain for each natural number n a continuous mapping g_n from E to $[-\frac{1}{3} \cdot (\frac{2}{3})^n, \frac{1}{3} \cdot (\frac{2}{3})^n]$ and a continuous mapping f_n from F to $[-(\frac{2}{3})^n, (\frac{2}{3})^n]$ with $f_{n+1} = (f_n - g_n)|F$ for all natural numbers n.

Consider the series of functions $\sum g_n$; we claim that this is uniformly convergent on E.

So let ε be any positive real number. There exists a positive integer N such that $(\frac{2}{3})^N < \varepsilon$. Then for every pair of natural numbers m, n such that $n > m > N$ we have

$$|g_m(x) + g_{m+1}(x) + \ldots + g_n(x)|$$
$$\leq |g_m(x)| + |g_{m+1}(x)| + \ldots + |g_n(x)|$$
$$\leq \frac{1}{3}(\frac{2}{3})^m + \frac{1}{3}(\frac{2}{3})^{m+1} + \ldots + \frac{1}{3}(\frac{2}{3})^n$$
$$\leq (\frac{2}{3})^m$$
$$< (\frac{2}{3})^N < \varepsilon$$

for all points x of E.

So $\sum g_n$ is uniformly convergent on E; if g is its sum function then g is continuous. We claim that $g(x) = f(x)$ for all points x of F. Let ε be any positive real number, x any point of F. Then for all natural numbers $m > N$ (with N as above) we have

$$|f(x) - g_0(x) - g_1(x) - \ldots - g_m(x)| = |f_{m+1}(x)| \leq (\frac{2}{3})^{m+1} < \varepsilon.$$

Thus $\sum g_n(x)$ converges to $f(x)$, i.e. $g(x) = f(x)$, as required.

169. Let A and B be disjoint T-closed subsets of E.

For $i = 1, 2$ the sets $A_i = A \cap F_i$ and $B_i = B \cap F_i$ are disjoint T_i-closed subsets of F_i. Hence for $i = 1, 2$ there are T-open sets U_i,

V_i such that $A_i \subseteq U_i \cap F_i$, $B_i \subseteq V_i \cap F_i$ and $U_i \cap V_i \cap F_i = \emptyset$. For $i = 1, 2$ let $U_i' = U_i \cup C_E F_i$ and $V_i' = V_i \cup C_E F_i$. Then for $i = 1, 2$ we see that U_i' and V_i' are T-open sets including A and B respectively. Let $U = U_1' \cap U_2'$, $V = V_1' \cap V_2'$. These are T-open sets which include A and B respectively. We claim that U and V are disjoint.

Suppose $x \in U \cap V$. Since $E = F_1 \cup F_2$ we have either $x \in F_1$ or $x \in F_2$, say $x \in F_1$. Then we have

$$x \in U \implies x \in U_1' \implies x \in U_1 \cup C_E F_1 \implies x \in U_1.$$

Hence $x \in U_1 \cap F_1$. Similarly we have $x \in V_1 \cap F_1$, which is a contradiction since $U_1 \cap V_1 \cap F_1 = \emptyset$. Thus (E, T) is normal.

170. Let π_1 and π_2 be the first and second coordinate maps from I^2 to I. Then $\pi_1 \circ f$ and $\pi_2 \circ f$ are continuous maps from F to I. So, by Tietze's Extension Theorem there are continuous maps g_1 and g_2 from E to I such that $g_i \,|\, F = \pi_i \circ f$ $(i = 1, 2)$. Define $g : E \to I^2$ by setting $g(x) = (g_1(x), g_2(x))$ for all x in E. Then g is continuous and $g \,|\, F = f$.

171. Let A and B be separated subsets of E.

Let a be any point of A. Since $a \in C_E \operatorname{Cl}_{T_d}(B)$, which is T_d-open, there exists a positive real number $r(a)$ such that $V_d(a, r(a)) \subseteq C_E \operatorname{Cl}_{T_d}(B)$. Let $U = \bigcup_{a \in A} V_d(a, \frac{1}{2} r(a))$. Then U is a T_d-open set which includes A. Similarly, if b is any point of B there is a positive real number $s(b)$ such that $V_d(b, s(b)) \subseteq C_E \operatorname{Cl}_{T_d}(A)$. Then $V = \bigcup_{b \in B} V_d(b, \frac{1}{2} s(b))$ is a T_d-open set which includes B. We claim that $U \cap V = \emptyset$.

For suppose there were a point x in $U \cap V$. Then there would be points a of A and b of B such that $x \in V_d(a, \frac{1}{2} r(a))$ and $x \in V_d(b, \frac{1}{2} s(b))$. Hence we would have $d(a, b) \leq d(a, x) + d(x, b) < \frac{1}{2} r(a) + \frac{1}{2} s(b) \leq \max \{r(a), s(b)\}$. So either $a \in V_d(b, s(b))$ or $b \in V_d(a, r(a))$, each of which is a contradiction. So (E, T) is completely normal.

172. (1) Suppose (E, T) is completely normal.

We shall prove that every subspace is not only normal but actually completely normal. Let F be any subset of E, A and B be separated subsets of F. Then we have

$$A \cap \operatorname{Cl}_T B = (A \cap F) \cap \operatorname{Cl}_T B = A \cap (F \cap \operatorname{Cl}_T B) = A \cap \operatorname{Cl}_{T_F} B = \emptyset.$$

Similarly $B \cap \text{Cl}_T A = \emptyset$. So A and B are separated subsets of E. Hence there are disjoint T-open subsets U and V of E such that $A \subseteq U$, $B \subseteq V$. Then $U \cap F$ and $V \cap F$ are disjoint T_F-open subsets of F such that $A \subseteq U \cap F$ and $B \subseteq V \cap F$. Thus (F, T_F) is completely normal.

(2) Conversely, suppose every subspace of (E, T) is normal.

Let A and B be separated subsets of E. Let $X = C_E(\text{Cl}_T A \cap \text{Cl}_T B)$. Then $X \cap \text{Cl}_T A$ and $X \cap \text{Cl}_T B$ are T_X-closed subsets of X and they are clearly disjoint. Since (X, T_X) is normal there are disjoint T_X-open sets U and V including $X \cap \text{Cl}_T A$ and $X \cap \text{Cl}_T B$ respectively. Since X is T-open, the sets U and V are actually T-open. Further

$$A = A \cap C_E \text{Cl}_T B \subseteq \text{Cl}_T A \cap C_E \text{Cl}_T B = \text{Cl}_T A \cap X \subseteq U.$$

Similarly $B \subseteq V$. So (E, T) is completely normal.

173. Denote the ordinary Euclidean topology on E by T^*.

Then all T^*-open sets are T-open. Since T^* is a T_1 topology it follows that T is a T_1 topology.

Now let A and B be disjoint T-closed sets. Let $A_1 = A \cap \mathbf{Q}$, $B_1 = B \cap \mathbf{Q}$. We claim that A_1 and B_1 are T^*-separated subsets of E, i.e. $\text{Cl}_{T^*}(A_1) \cap B_1 = \emptyset = \text{Cl}_{T^*}(B_1) \cap A_1$. Suppose we have a point x in $\text{Cl}_{T^*} A_1 \cap B_1$. Let U be any T-open set containing x. Then $U = U^* \cup V$ where $U^* \in T^*$ and $V \subseteq C_E \mathbf{Q}$. Since $x \in B_1 \subseteq \mathbf{Q}$ we must have $x \in U^*$. Hence U^* meets A_1 (because $x \in \text{Cl}_{T^*} A_1$). So U meets A. Thus $x \in \text{Cl}_T A \cap B = A \cap B = \emptyset$, which is a contradiction. So $\text{Cl}_{T^*} A_1 \cap B_1 = \emptyset$ and similarly $\text{Cl}_{T^*} B_1 \cap A_1 = \emptyset$.

Thus A_1 and B_1 are separated subsets of E with its usual topology T^* which is induced by a metric and so completely normal. Thus there are disjoint T^*-open subsets U_1, V_1 of E such that $U_1 \supseteq A_1$, $V_1 \supseteq B_1$. Set $U = U_1 \cup (A \cap (C_E \mathbf{Q}))$, $V = V_1 \cup (B \cap C_E \mathbf{Q})$. Then U and V are T-open disjoint sets which include A and B respectively.

So (E, T) is normal and hence T_4.

174. The implications $T_1 \implies T_0$, $T_2 \implies T_1$, $T_{2\frac{1}{2}} \implies T_2$ are immediate consequences of the definitions.

To show that $T_3 \implies T_{2\frac{1}{2}}$ let x and y be distinct points of E. Since T_3 spaces are T_1, it follows that $\{y\}$ is a T-closed set. Since $x \notin \{y\}$ it follows from the regularity of T that there are disjoint T-open sets U_1, V_1 such that $x \in U_1$ and $\{y\} \subseteq V_1$, i.e. $y \in V_1$. Now $\text{Cl}_T V_1$ is a T-closed set and x does not belong to it since it has a T-neighbourhood (U_1) which does not meet V_1. Hence, using the regularity again, we obtain disjoint T-open sets U_2, V_2 such that $x \in U_2$ and $\mathrm{Cl}_T V_1 \subseteq V_2$. We claim that $\mathrm{Cl}_T U_2 \cap \mathrm{Cl}_T V_1 = \emptyset$. If there were a point z in this intersection, we would have $z \in V_2$; but then V_2 would be a neighbourhood of z which does not meet U_2, contradicting the hypothesis that $z \in \mathrm{Cl}\, U_2$. So U_2 and V_1 are T-neighbourhoods of x and y respectively whose closures are disjoint. Thus T is completely Hausdorff.

Suppose T is $T_{3\frac{1}{2}}$. Then T is T_1 and completely regular. Hence, using Exercise 162, we see that T is T_1 and regular, i.e. T_3.

Suppose T is T_4. Then T is certainly T_1. To show that T is completely regular let a be a point of E and F a T-closed subset of E not containing a. Then $\{a\}$ and F are disjoint T-closed subsets (using the T_1 property). Urysohn's Lemma provides the continuous mapping from E to $[0,1]$ required to show that T is completely regular. So T is $T_{3\frac{1}{2}}$.

Finally, if T is T_5 then it is T_1 and completely normal. Since disjoint closed subsets are certainly separated it follows from the definition of complete normality that T is normal and hence T_4.

Chapter 13

ANSWERS FOR CHAPTER 6

175. Let E be an infinite set, T the finite complement topology on E.

Let $(U_i)_{i \in I}$ be a T-open cover of E. Let U_{i_0} be a non-empty set of the cover. Then $C_E(U_{i_0})$ is finite; say $C_E(U_{i_0}) = \{p_1, \ldots, p_n\}$. For $j = 1, \ldots, n$ there is a set U_{i_j} of the cover such that $p_j \in U_{i_j}$. Then $E = U_{i_0} \cup U_{i_1} \cup \ldots \cup U_{i_n}$. So (E, T) is compact.

176. (1) Every topological space (E, T) in which E is a finite set is compact. So in particular every finite discrete space is compact.

(2) Let (E, T) be a discrete space in which E is infinite.

Then $(\{x\})_{x \in E}$ is an open cover of E which has no finite subcover. So (E, T) is not compact.

177. (1) Suppose A is compact.

Let $(F_i)_{i \in I}$ be a family of T_A-closed subsets of A with the finite intersection property. For each index i in I there is a T-closed subset F_i' of E such that $F_i = F_i' \cap A$. If $\bigcap_{i \in I} F_i = \emptyset$ we have $A \cap (\bigcap_{i \in I} F_i') = \emptyset$ and hence $A \subseteq C_E(\bigcap_{i \in I} F_i') = \bigcup_{i \in I} C_E(F_i')$. So $(C_E(F_i'))_{i \in I}$ is a T-open cover of A. By hypothesis there is a finite subset J of I such that $\bigcup_{i \in J} C_E(F_i') \supseteq A$, whence $C_E(\bigcap_{i \in J} F_i') \supseteq A$ and so $\bigcap_{i \in J} (A \cap F_i') = \emptyset$, i.e. $\bigcap_{i \in J} F_i = \emptyset$. This is a contradiction (since (F_i) has the finite intersection property). Hence $\bigcap_{i \in I} F_i \neq \emptyset$.

(2) Conversely, suppose the condition is satisfied.

Let $(U_i)_{i \in I}$ be a T-open cover of A. Then $\bigcup_{i \in I} U_i \supseteq A$ and so $A \cap C_E(\bigcup_{i \in I} U_i) = \emptyset$. Thus $(A \cap C_E(U_i))_{i \in I}$ is a family of T_A-closed subsets of A such that the total intersection is empty. Hence this family cannot have the finite intersection property. So there is a finite subset J of I such that $\bigcap_{i \in J} (A \cap C_E(U_i)) = \emptyset$.

Hence $A \cap (\bigcap_{i \in J} C_E(U_i)) = A \cap C_E(\bigcup_{i \in J} U_i) = \emptyset$ and so we have $A \subseteq \bigcup_{i \in J} U_i$.

Hence A is compact.

178. (1) $\Longrightarrow$ (2) Suppose (E, T) is compact.

Let F be a filter on E. Then the family $(\mathrm{Cl}X)_{X \in F}$ has the finite intersection property. Hence $\bigcap_{X \in F} \mathrm{Cl}X$ is non-empty, i.e. F has at least one adherent point.

(2) $\Longrightarrow$ (3) Suppose every filter on E has at least one adherent point.

Let F be an ultrafilter on E. Then F has an adherent point p, say. But every adherent point of an ultrafilter is a limit point. So F converges to p.

(3) $\Longrightarrow$ (1) Suppose every ultrafilter on E converges.

Let $(F_i)_{i \in I}$ be a family of T-closed subsets of E with the finite intersection property. This family generates a filter which is in turn included in an ultrafilter Φ, which converges to a point p, say. We claim that $p \in \bigcap_{i \in I} F_i$. So let F_i be any set of the family. For every T-neighbourhood V of p we have $F_i \in \Phi$ and $V \in \Phi$. So $F_i \cap V \neq \emptyset$. Thus $p \in \mathrm{Cl}F_i = F_i$. So $\bigcap_{i \in I} F_i \neq \emptyset$ and hence (E, T) is compact.

179. Let (E, T) be a Hausdorff space, K a compact subset of E.

Let a be a point of $C_E(K)$. For each point x of K there are disjoint T-open sets U_x, V_x such that $x \in U_x$ and $a \in V_x$. Then $(U_x)_{x \in K}$ is an open cover of K. So there is a finite subcover, that is to say there is a finite subset $\{x_1, \ldots, x_n\}$ of K such that $K \subseteq U_{x_1} \cup U_{x_2} \cup \ldots \cup U_{x_n}$. Let $V = V_{x_1} \cap V_{x_2} \cap \ldots \cap V_{x_n}$; V is an open set containing a and $V \cap (U_{x_1} \cup \ldots \cup U_{x_n}) = \emptyset$. Hence $V \subseteq C_E(K)$. So $C_E(K)$ is a T-neighbourhood of each of its points, hence T-open.

So K is T-closed.

180. Let $(K_i)_{i \in I}$ be a finite family of compact subsets of E.

Let $(U_j)_{j \in J}$ be an open cover of $\bigcup_{i \in I} K_i$. Then of course $(U_j)_{j \in J}$ is an open cover of each compact set K_i. So for each index i in I there is a finite subset J_i of J such that $\bigcup_{j \in J_i} U_j \supseteq K_i$. Let $J' = \bigcup_{i \in I} J_i$; J' is a finite set and $\bigcup_{j \in J'} U_j \supseteq \bigcup_{i \in I} K_i$. So $\bigcup_{i \in I} K_i$ is compact.

181. Let (E, T) be a compact space, F a T-closed subset of E.

Let $(U_i)_{i \in I}$ be a T-open cover of F, so that $\bigcup_{i \in I} U_i \supseteq F$. Then $C_E(F) \cup (\bigcup_{i \in I} U_i) = E$. Hence there is a finite subset J of I such that $C_E(F) \cup (\bigcup_{i \in J} U_i) = E$. Then $\bigcup_{i \in J} U_i \supseteq F$. So F is compact.

182. Let $(K_i)_{i \in I}$ be a family of compact subsets of the Hausdorff space (E, T).

Then each of the sets K_i is T-closed. So $\bigcap_{i \in I} K_i$ is T-closed and is a subset of each of the compact sets K_i. Hence $\bigcap_{i \in I} K_i$ is compact.

183. Let (E, T) be a compact Hausdorff space. Let A and B be disjoint closed subsets of E.

Then A is compact, and as in Exercise 179 we can produce for every point b of B an open set U_b which includes A and an open set V_b containing b such that $U_b \cap V_b = \emptyset$. Then $(V_b)_{b \in B}$ is an open cover of B, which is also compact. So there is a finite subset $\{b_1, \ldots, b_n\}$ of B such that $B \subseteq V_{b_1} \cup \ldots \cup V_{b_n} = V$, say. Let $U = U_{b_1} \cap \ldots \cap U_{b_n}$. Then U and V are disjoint open sets which include A and B respectively. So (E, T) is normal.

184. Let $(U_i')_{i \in I}$ be a T'-open cover of $f^{\rightarrow}(E)$.

Then $(f^{\leftarrow}(U_i'))_{i \in I}$ is a T-open cover of E (since f is (T, T')-continuous). Hence there is a finite subset J of I such that $E = \bigcup_{i \in J} f^{\leftarrow}(U_i')$. Then we have $f^{\rightarrow}(E) = f^{\rightarrow}(\bigcup_{i \in J} f^{\leftarrow}(U_i')) = \bigcup_{i \in J} f^{\rightarrow}(f^{\leftarrow}(U_i')) \subseteq \bigcup_{i \in J} U_i'$. So $f^{\rightarrow}(E)$ is compact.

185. Let (E, T) be compact, (E', T') Hausdorff and f a (T, T')-continuous bijection from E onto E'. We have to show that f^{-1} is (T', T)-continuous.

So let F be any T-closed subset of E. Since E is compact and F is closed it follows that F is compact. Since f is continuous, $f^{\rightarrow}(F)$ is compact. Since E' is Hausdorff, $f^{\rightarrow}(F)$ is closed, i.e. $(f^{-1})^{\leftarrow}(F)$ is

T'-closed. So f^{-1} is (T', T)-continuous.

186. (1) Suppose all the spaces (E_i, T_i) are compact. Let U be an ultrafilter on E.

For each index i in I the collection $\{\pi_i^\rightarrow(X)\}_{X \in U}$ is base for an ultra-filter U_i on E_i. Since each (E_i, T_i) is compact each of these ultrafilters U_i converges to a point a_i of E_i. Let a be the point of E such that $\pi_i(a) = a_i$ for each i in I. We claim that U converges to a. So let V be any T-neighbourhood of a. Then there exists a finite subset J of I and a family $(X_i)_{i \in J}$ such that for each i in J the set X_i is a T_i-open subset of E_i and $a \in \bigcap_{i \in J} \pi_i^\leftarrow(X_i) \subseteq V$. For each index i in J the set X_i is a T_i-neighbourhood of a_i and so belongs to U_i. Hence for each index i in J we have $\pi_i^\leftarrow(X_i) \in U$. Hence, since J is finite, we have $V \in U$.

Thus U converges to a. So (E, T) is compact.

(2) Conversely, suppose all the sets E_i are non-empty and (E, T) is compact.

For each index i in I the projection π_i is a (T, T_i)-continuous surjection. So for each index i the set $E_i = \pi_i^\rightarrow(E)$ is compact.

187. (1) $\implies$ (2) Compare Exercise 177.

(3) $\implies$ (4) Suppose every countably infinite subset has an ω-accumulation point.

Let $\sigma : \mathbf{N} \to E$ be a sequence of points in E; let $A = \sigma^\rightarrow(\mathbf{N})$. If A is finite then σ clearly has an adherent point. If A is infinite then it has an ω-accumulation point a, and a is clearly an adherent point of σ.

(4) $\implies$ (3) Suppose every sequence has an adherent point.

Let A be a countably infinite subset of E. Then there is a bijection σ from $\mathbf{N}$ onto A, which is of course a sequence in E. By hypothesis σ has an adherent point a in E. Then a is an ω-accumulation point of A.

(3) $\implies$ (1) Suppose every countably infinite subset has an ω-accumulation point.

Let $(U_n)_{n \in \mathbf{N}}$ be a countable open cover of E; we may assume that the sets U_n are all distinct and that none is included in the union of those that precede it. Suppose this cover has no finite subcover. Then for each natural number m there is a point x_m such that $x_m \notin \bigcup_{n < m} U_n$ and $x_m \in \bigcup_{n \leq m} U_n$. The points x_m are all distinct, so the set $X = \{x_m\}$ is infinite.

We claim that X has no ω-accumulation point. For if p is any point of E there is a natural number k such that $p \in U_k$ and U_k is a T-neighbourhood of p which contains only finitely many points of X.

(1) $\implies$ (3) Suppose there exists a countably infinite subset S of E with no ω-accumulation point.

Then for every point x of E there must be an open subset U_x containing x but at most finitely many points of S. Let Z be the (countable) set of finite subsets of S. For each set F in Z let $I_F = \{x \in E : U_x \cap S = F\}$ and $U_F = \bigcup_{x \in I_F} U_x$. Then $(U_F)_{F \in Z}$ is a countable open cover of E. But if $\{F_1, F_2, \ldots, F_k\}$ is any finite subset of Z we have
$$S \cap (U_{F_1} \cup U_{F_2} \cup \ldots \cup U_{F_k}) = (S \cap U_{F_1}) \cup \ldots \cup (S \cap U_{F_k}) = F_1 \cup \ldots \cup F_k,$$
which is finite. Thus the countable open cover $(U_F)_{F \in Z}$ has no finite subcover. So (E, T) is not countably compact.

188. Trivial.

189. Let (E, T) be a compact space, (E', T') a countably compact space. Let $(U_n)_{n \in \mathbb{N}}$ be a countable $(T \times T')$-open cover of $E \times E'$.

For each natural number n set $G_n = \bigcup_{0 \le k \le n} U_k$. Then $G_n \subseteq G_{n+1}$ for all natural numbers n. For each natural number n let H_n be the set of points y of E' for which there is a T'-neighbourhood V' of y such that $E \times V' \subseteq G_n$.

Each set H_n is T'-open, for if $y \in E'$ and V' is a T'-neighbourhood of y such that $E \times V' \subseteq G_n$ there is a T'-neighbourhood W' of y such that V' is a T'-neighbourhood of all points of W'. This shows that $W' \subseteq H_n$. Thus H_n is a T'-neighbourhood of each of its points y, and hence is T'-open.

Since $G_n \subseteq G_{n+1}$ we have $H_n \subseteq H_{n+1}$ for all natural numbers n. We claim that $\bigcup_{n \in \mathbb{N}} H_n = E'$.

So let b be any point of E'. For each point x of E the point (x, b) belongs to one of the sets in the family (G_n), say $(x, b) \in G_{n(x)}$. Thus for each point x of E there exists a T-open set V_x and a T'-open set V'_x such that $(x, b) \in V_x \times V'_x \subseteq G_{n(x)}$. So $B = E \times \{b\} \subseteq \bigcup_{x \in E} (V_x \times V'_x)$. But B is clearly homeomorphic to E and hence is compact. So there is a finite subset $\{x_1, \ldots, x_n\}$ of E such that $B \subseteq \bigcup_{1 \le k \le n} V_{x_k} \times V'_{x_k}$. Then $B \subseteq \bigcup_{1 \le k \le n} G_{n(x_k)} = G_{m(b)}$ where $m(b) = \max\{n(x_1), \ldots, n(x_n)\}$. Let $V' = \bigcap_{1 \le k \le n} V'_{x_k}$; then V' is a T'-neighbourhood of b. Then we have $B \subseteq (\bigcup_{1 \le k \le n} V_{x_k}) \times V' = E \times V' \subseteq G_{m(b)}$ and hence $b \in H_{m(b)}$.

Thus $(H_n)_{n \in \mathbb{N}}$ is a countable T'-open cover for E'. So there is a natural number p such that $H_p = E'$. Then for every point y of E' there is a T'-neighbourhood V' such that $E \times V' \subseteq G_p$, i.e. for every point x of E we have $(x, y) \in G_p$. Thus $G_p = E \times E'$.

So $E \times E' = U_1 \cup U_2 \cup \ldots \cup U_p$.

190. (1) Countably compact: Compare Exercise 184.

(2) Sequentially compact: Let σ be a sequence in $f^{\rightarrow}(E)$, where E is sequentially compact. For each natural number n let x_n be a point of E such that $f(x_n) = \sigma(n)$. Thus we have a sequence ν in E such that $\sigma = f \circ \nu$ (ν is defined by setting $\nu(n) = x_n$). By hypothesis ν has a convergent subsequence ν'. Then $f \circ \nu'$ is a convergent subsequence of σ.

191. Let γ be an ordinal number. Let $\Gamma = \gamma \cup \{\gamma\}$. Then Γ is well-ordered. Every non-empty subset X of Γ has a greatest lower bound (its first member) and a least upper bound (the first member of the set of all upper bounds of X, which is non-empty since γ is an upper bound of X).

Let $(U_i)_{i \in I}$ be an open cover of Γ (where Γ has the order topology). Let $S = \{y \in \Gamma : [0, y) \text{ can be covered by finitely many members of the cover}\}$. Let α be the least upper bound of S. There is an index i_0 in I such that $\alpha \in U_{i_0}$. The set U_{i_0} is included in S. If $\alpha \neq \gamma$ there is an interval (x, y) such that $\alpha \in (x, y) \subseteq U_{i_0}$; then $y \in S$ but $y > \alpha$, which is a contradiction. So $\alpha = \gamma$, and hence $S = \Gamma$.

Thus Γ is compact.

In particular $\Omega_1 = \omega_1 \cup \{\omega_1\}$ is compact and hence countably compact. Thus every infinite subset of ω_1 has an ω-acculumation point in Ω_1. But ω_1 is not an ω-accumulation point of any infinite subset of ω_1. So every infinite subset of ω_1 has an ω-accumulation point in ω_1.

Thus ω_1 is countably compact.

To see that ω_1 is sequentially compact let σ be a sequence in ω_1.

If $\sigma^{\rightarrow}(\mathbf{N})$ is finite then σ clearly has a convergent subsequence. So suppose $\sigma^{\rightarrow}(\mathbf{N})$ is infinite. Then $\sigma^{\rightarrow}(\mathbf{N})$ has an ω-accumulation point, α say, in ω_1. α has a countable neighbourhood base (consisting of $\{\alpha\}$ alone if α is not a limit ordinal and of the countably many intervals $(\beta, \alpha]$ where β is an ordinal less than α if α is a limit ordinal) and so there is a sequence $(V_n)_{n \in \mathbf{N}}$ of neighbourhoods of α such that $V_n \subseteq V_{n+1}$ for all natural numbers n. By choosing a point of $\sigma^{\rightarrow}(\mathbf{N})$ from each neighbourhood V_n we obtain a subsequence which converges to α.

So ω_1 is sequentially compact. But ω_1 is not compact since the cover $([0, \alpha))_{\alpha \in \omega_1}$ has no finite subcover.

192. I is compact by the Heine-Borel Theorem; hence P is compact by Tihonov's Theorem.

(According to the Heine-Borel Theorem of real analysis we know that every bounded closed inverval $E = [a, b]$ is compact. To prove this let $(U_i)_{i \in I}$ be an open cover of E and let A be the set of points x of E such that the subinterval $[a, x]$ can be covered by finitely many sets of the cover; then let $c = \sup E$ and prove that $c = b$ and $c \in A$.)

The elements of P are functions from I to I. We claim that a sequence (f_n) of elements of P converges to the point f of P if and only if the sequence $(f_n(x))$ of points of I converges to $f(x)$ for all points x of I.

Suppose first that (f_n) converges to f and let x be any point of I. Let V be any neighbourhood of $f(x)$ in I. Let $V_t = I$ for $t \neq x$ and let $V_x = V$. Then $\prod_{t \in I} V_t$ is a neighbourhood of f in P. There is thus a positive integer n_0 such that for all $n > n_0$ we have $f \in \prod_{t \in I} V_t$. Hence for all $n > n_0$ we have $f_n(x) \in V_x = V$. So $(f_n(x))$ converges to $f(x)$.

Conversely, suppose (f_n) is a sequence of elements of P such that $(f_n(x))$ converges to $f(x)$ for every point x of I.

Let V be any neighbourhood of f in P. Then there is a family $(X_t)_{t \in I}$ of open subsets of I, and a finite subset J of I such that $X_t = I$ for all t not in J and $f \in \prod_{t \in I} X_t \subseteq V$. For each x in J the sequence $(f_n(x))$ converges to $f(x)$. So there is, for each x in J, a positive integer n_x such that for all $n > n_x$ we have $f_n(x) \in X_x$. Let $n_0 = \max_{x \in J}\{n_x\}$. Then for all $n > n_0$ we have $f_n \in \prod_{t \in I} X_t$. Thus (f_n) converges to f.

Now we exhibit a sequence in P which has no convergent subsequence.

Let (f_n) be given by setting

$$f_n(x) = \text{the } n\text{th digit in the binary expansion of } x.$$

Suppose (f_{n_k}) were a convergent subsequence. Let p be the point of I such that $f_{n_k}(p) = 0$ or 1 according as n_k is odd or even. The sequence $(f_{n_k}(p))$ is $0, 1, 0, 1, \ldots$, which does not converge. So (f_{n_k}) does not converge. Thus P is not sequentially compact.

193. By Exercise 191 ω_1 is countably compact; P is compact by Tihonov's Theorem. So $\omega_1 \times P$ is countably compact by Exercise 189.

If $\omega_1 \times P$ were compact then each factor would be compact (since each continuous image of a compact spact is compact). But ω_1 is not compact (Exercise 191).

If it were sequentially compact then each factor would be sequentially compact (since each continuous image of a sequentially compact space is sequentially compact (Exercise 190)). But P is not sequentially compact (Exercise 192).

So $\omega_1 \times P$ is neither compact nor sequentially compact.

194. Let $(U_i)_{i \in I}$ be an open cover of E.

For at least one index i_0 in I we have $p \in U_{i_0}$. Since $p \notin C_E(U_{i_0})$ it follows that $C_E(U_{i_0})$ is finite, consisting of points $x_1, \ldots, x_n$ say. For $k = 1, \ldots, n$ there is an index i_k in I such that $x_{i_k} \in U_{i_k}$. Then $E = U_{i_0} \cup U_{i_1} \cup \ldots \cup U_{i_n}$. So E is compact.

Let σ be a sequence in E. If $\sigma^{\rightarrow}(\mathbf{N})$ is finite then σ has a convergent subsequence. So suppose $\sigma^{\rightarrow}(\mathbf{N})$ is infinite; in this case σ has a subsequence converging to p.

195. Let $E_1 = E \cup \{\infty\}$ where $\infty \notin E$. Let $T_1 = T \cup T_0$ where T_0 is the collection of all subsets of E_1 of the form $\{\infty\} \cup C_E(K)$ where K is a compact subset of E.

We claim that T_1 is a topology on E_1.

(1) $\emptyset \in T$, so $\emptyset \in T_1$.

(2) $E_1 = \{\infty\} \cup C_E(\emptyset)$ and $\emptyset$ is compact; so $E_1 \in T_1$.

(3) Since T is a topology the union of each family of sets in T belongs to T, hence to T_1.

Let $(\{\infty\} \cup C_E(K_i))_{i \in I}$ be a family of sets in T_0, where each K_i is compact. Then $\bigcup_{i \in I} (\{\infty\} \cup C_E(K_i)) = \{\infty\} \cup (\bigcup_{i \in I} C_E(K_i)) = \{\infty\} \cup C_E(\bigcap_{i \in I} K_i)$, which belongs to T_0 since $\bigcap_{i \in I} K_i$ is compact (see Exercise 180). It follows that $\bigcup_{i \in I} (\{\infty\} \cup C_E(K_i))$ is in T_1.

Finally let $U \in T$ and $\{\infty\} \cup C_E(K) \in T_0$, where K is compact. Then we have $U \cup (\{\infty\} \cup C_E(K)) = \{\infty\} \cup C_E(C_E(U)) \cup C_E(K) = \{\infty\} \cup C_E(K \cap C_E(U))$. Now $K \cap C_E(U)$ is compact. It follows that $U \cup (\{\infty\} \cup C_E(K)) \in T_0$ and hence belongs to T_1.

So T_1 is closed under the formation of unions.

(4) Since T is a topology the intersection of any finite family of sets

in T is again in T.

Let $(\{\infty\} \cup C_E(K_i))_{i \in I}$ be a finite family of sets in T_0 (where the sets K_i are compact). Then $\bigcap_{i \in I} (\{\infty\} \cup C_E(K_i)) = \{\infty\} \cup (\bigcap_{i \in I} C_E(K_i))$ $= \{\infty\} \cup C_E(\bigcup_{i \in I} K_i)$. Since $\bigcup_{i \in I} K_i$ is compact (Exercise 181) we see that $\bigcap_{i \in I} (\{\infty\} \cup C_E(K_i))$ belongs to T_0, hence to T_1.

Finally let U be a set in T, and let $\{\infty\} \cup C_E(K)$, with K compact, be a set in T_0. Then $U \cap (\{\infty\} \cup C_E(K)) = U \cap C_E(K)$, which belongs to T since K is T-closed (Exercise 179). Hence $U \cap (\{\infty\} \cup C_E(K))$ is in T_1.

So T_1 is closed under the formation of finite intersections.

Thus T_1 is a topology.

Clearly we have an injection $i : E \to E_1$ and $(T_1)_{i \to (E)} = T$; so i is a homeomorphism from E onto $i^\to(E)$.

Now we claim that T_1 is Hausdorff.

Let a and b be distinct points of E_1. If a and b both belong to E then they have disjoint T-neighbourhoods, which are also T_1-neighbourhoods. If $a \in E$ and $b = \infty$ then, since (E, T) is locally compact, a has a compact T-neighbourhood K. Then K and $\{\infty\} \cup C_E(K)$ are disjoint T_1-neighbourhoods of a and $b = \infty$.

Finally we show that (E_1, T_1) is compact.

Let $(V_i)_{i \in I}$ be a T_1-open cover of E_1. There is at least one index i_0 in I such that $\infty \in V_{i_0}$; say $V_{i_0} = \{\infty\} \cup C_E(K)$, where K is compact. The family $(V_i \cap E)_{i \in I}$ is a T-open cover of K. Hence there is a finite set $\{i_1, \ldots, i_n\}$ of indices in I such that $K \subseteq (V_{i_1} \cap E) \cup \ldots \cup (V_{i_n} \cap E)$. It follows at once that $E_1 = V_{i_0} \cup V_{i_1} \cup \ldots \cup V_{i_n}$. Thus E_1 is compact.

196. Since (E', T') is a compact Hausdorff space it is normal and hence completely regular.

Let $P_{E'} = \prod_{f' \in C^*(E')} I_{f'}$, where $I_{f'} = [0, 1]$ for each member f' of $C^*(E')$. Let $e' : E' \to P_{E'}$ be given by $(e'(x'))_{f'} = f'(x')$ for all x' in E' and all f' in $C^*(E')$. For each mapping f' in $C^*(E')$ the mapping $\pi_{f' \circ f} : P_E \to I$ is continuous. Hence there exists a continuous mapping H from P_E to $P_{E'}$ such that $\pi'_{f'} \circ H = \pi_{f' \circ f}$ for each f' in $C^*(E')$. Then for each f' in $C^*(E')$ we have

$$\pi'_{f'} \circ H \circ e = \pi_{f' \circ f} \circ e = f' \circ f = \pi_{f'} \circ e' \circ f$$

and so $H \circ e = e' \circ f$.

Now $(e')^\to(E')$, being the image of a compact set E' under a continuous mapping, is a compact subset of the Hausdorff space $P_{E'}$ and hence

is closed. So we have $H^{\to}(\beta E) = H^{\to}(\mathrm{Cl}(e^{\to}(E))) \subseteq \mathrm{Cl}(H^{\to}(e^{\to}(E)))$
$= \mathrm{Cl}((e')^{\to}(f^{\to}(E))) \subseteq \mathrm{Cl}((e')^{\to}(E')) = (e')^{\to}(E')$.

Let $\beta f = (e')^{-1} \circ (H \,|\, \beta E)$; then βf is a continuous mapping from βE to E' and we have

$$(\beta f) \circ e = (e')^{-1} \circ (H \,|\, \beta E) \circ e = (e')^{-1} \circ e' \circ f = f.$$

Chapter 14

ANSWERS FOR CHAPTER 7

197. If A and B are disjoint closed sets we have $A \cap \text{Cl } B = A \cap B = \emptyset$ and $B \cap \text{Cl } A = B \cap A = \emptyset$. So A and B are separated.

If A and B are disjoint open sets we have $A \subseteq C_E(B)$, which is closed. So $\text{Cl } A \subseteq C_E(B)$ and hence $\text{Cl } A \cap B = \emptyset$. Similarly $A \cap \text{Cl } B = \emptyset$. So again A and B are separated.

198. (1) $\implies$ (2) Suppose (E, T) is disconnected.

Then there is a subset A of E, $A \neq \emptyset$, $A \neq E$, which is both T-open and T-closed. Then A and $C_E(A)$ are disjoint T-closed, hence separated, sets such that $E = A \cup C_E(A)$.

(2) $\implies$ (3) Suppose $E = A \cup B$ where A and B are non-empty separated sets. Then

$$\text{Cl } A = \text{Cl } A \cap (A \cup B) = (\text{Cl } A \cap A) \cup (\text{Cl } A \cap B) = \text{Cl } A \cap A = A.$$

So A is closed. Similarly B is closed.

(3) $\implies$ (4) Suppose $E = A \cup B$ where A and B are disjoint non-empty closed sets.

Then $A = C_E(B)$ and $B = C_E(A)$ are disjoint non-empty open sets whose union is E.

(4) $\implies$ (1) Suppose $E = A \cup B$ where A and B are disjoint non-empty open sets.

Then $A = C_E(B)$ and hence is closed. So (E, T) is disconnected.

199. (1) Suppose A is disconnected.

Then $A = X \cup Y$ where X and Y are disjoint non-empty T_A-closed sets. We claim that X and Y are separated subsets of E. This follows because $X \cap \mathrm{Cl}_T Y = X \cap A \cap \mathrm{Cl}_T Y = X \cap \mathrm{Cl}_{T_A} Y = X \cap Y = \emptyset$ and similarly $Y \cap \mathrm{Cl}_{T_A} X = \emptyset$.

(2) Suppose $A = X \cup Y$ where X and Y are non-empty separated subsets of E.

We have $X \cap \mathrm{Cl}_{T_A} Y = X \cap A \cap \mathrm{Cl}_T Y = X \cap \mathrm{Cl}_T Y = \emptyset$ and similarly $Y \cap \mathrm{Cl}_{T_A} X = \emptyset$. So A is the union of two non-empty separated subsets of A. Thus A is disconnected.

200. Suppose A is connected and $A \subseteq B \subseteq \mathrm{Cl}\, A$.

If B is disconnected we have $B = X \cup Y$ where X and Y are non-empty separated subsets of E. Then $A = (A \cap X) \cup (A \cap Y)$ and $A \cap X$ and $A \cap Y$ are separated. Since A is connected we have either $A \cap X = \emptyset$ or $A \cap Y = \emptyset$, say $A \cap X = \emptyset$. So $A \subseteq Y$. It follows that $X \subseteq B \subseteq \mathrm{Cl}_T A \subseteq \mathrm{Cl}_T Y$, whence $X = X \cap \mathrm{Cl}_T Y = \emptyset$. This is a contradiction. So B is connected.

201. Suppose $A = \bigcup_{i \in I} A_i$ is disconnected; say $A = X \cup Y$ where X and Y are non-empty separated subsets of E.

For each index i in I we have either $A_i \cap X = \emptyset$ or $A_i \cap Y = \emptyset$. Suppose that for some index i_0 in I we have $A_{i_0} \cap X = \emptyset$. Then $A_{i_0} \subseteq Y$. Hence $\bigcap_{i \in I} A_i \subseteq Y$. It follows that for every index i in I we have $A_i \cap X = \emptyset$. Thus $X = X \cap (\bigcup_{i \in I} A_i) = \emptyset$, which is a contradiction. So A is connected.

202. Suppose $A \cup B = X \cup Y$ where X and Y are separated sets.

Then $A = (A \cap X) \cup (A \cap Y)$ and $A \cap X$ and $A \cap Y$ are separated. Hence either $A \cap X = \emptyset$ or $A \cap Y = \emptyset$, say $A \cap Y = \emptyset$ and so $A \subseteq X$. Similarly we have either $B \subseteq X$ or $B \subseteq Y$. If $B \subseteq Y$ we have $\emptyset \neq A \cap \mathrm{Cl}\, B \subseteq X \cap \mathrm{Cl}\, Y = \emptyset$, which is a contradiction. So $B \subseteq X$. But then we have $A \cup B \subseteq X$ and so $Y = \emptyset$, which is again a contradiction.

203. For each natural number n let $B_n = A_0 \cup A_1 \cup \ldots \cup A_n$.

Then clearly $\bigcup_{n\in\mathbf{N}} B_n = \bigcup_{n\in\mathbf{N}} A_n$. Now $\bigcap_{n\in\mathbf{N}} B_n = B_0 = A_0$ which is non-empty, since $A_0 \cap A_1 \neq \emptyset$. Each set B_n is connected (by a simple inductive argument). It follows that $\bigcup_{n\in\mathbf{N}} A_n = \bigcup_{n\in\mathbf{N}} B_n$ is connected (by Exercise 201).

204. Suppose that $\mathbf{Z} = A \cup B$ where A and B are separated and A is non-empty.

Let k be an element of A.

If k is even then $V = \{k-1, k, k+1\}$ is a T-neighbourhood of k. Since $k \notin \mathrm{Cl}_T B$ it follows that V does not meet B; so $k-1$ and $k+1$ are not in B. Hence $k-1$ and $k+1$ are in A.

If k is odd then $W = \{k, k+1, k+2\}$ is the smallest T-neighbourhood of $k+1$; since it meets A it follows that $k+1 \in \mathrm{Cl}_T A$ and hence that $k+1 \notin B$; hence $k+1 \in A$. Similarly by considering the T-neighbourhood $\{k-2, k-1, k\}$ of $k-1$ we deduce that $k-1 \in A$.

It follows by induction (upwards and downwards) that every integer is in A. Thus $A = \mathbf{Z}$, $B = \emptyset$ and $(\mathbf{Z}, T)$ is connected.

205. (1) Let A be a subset of $\mathbf{R}$ which is not an interval. Then there must be points a, b of A with $a < b$ such that the interval $[a, b]$ is not included in A; so there must be a point c such that $a < c < b$ and $c \notin A$. Then the intersections of A with the intervals $(-\infty, c)$ and (c, ∞) form a disconnection of A.

So a connected subset of $\mathbf{R}$ must be an interval.

(2) Conversely, let E be an interval of $\mathbf{R}$.

Suppose E is disconnected; then there are closed subsets F and G of $\mathbf{R}$ such that $E = (E \cap F) \cup (E \cap G)$, $E \cap F \neq \emptyset$, $E \cap G \neq \emptyset$ and $E \cap F \cap G = \emptyset$. Write $A = E \cap F$, $B = E \cap G$. Let a and b be points of A and B respectively; we may assume that $a < b$. Then $[a, b] \subseteq E$.

The set $X = [a, b] \cap A$ is non-empty and is bounded above by b; so it has a least upper bound, say c. Since $a \leq c \leq b$ and a and b are in the interval E it follows that c is in E. Since c is the least upper bound of X it follows that for every positive real number ε there is a point of X greater than $c - \varepsilon$: this means that every neighbourhood of c meets X; so c is in the closure of X and hence in A. Again, since c is an upper bound of X it follows that for every positive real number ε such that $c + \varepsilon < b$ we have $c + \varepsilon \notin X$, hence $c + \varepsilon \in B$. It follows that c is in the closure of B, hence in B.

But this is a contradiction since A and B are disjoint. Thus E is connected.

206. Suppose $f^{\rightarrow}(A) = (f^{\rightarrow}(A) \cap X') \cup (f^{\rightarrow}(A) \cap Y')$ where X' and Y' are T'-open subsets of E' such that $f^{\rightarrow}(A) \cap X' \cap Y' = \emptyset$, but $f^{\rightarrow}(A) \cap X' \neq \emptyset$ and $f^{\rightarrow}(A) \cap Y' \neq \emptyset$. Since f is (T, T')-continuous, $f^{\leftarrow}(X')$ and $f^{\leftarrow}(Y')$ are T-open. Further, $A \cap f^{\leftarrow}(X')$ and $A \cap f^{\leftarrow}(Y')$ are non-empty and $A \cap f^{\leftarrow}(X') \cap f^{\leftarrow}(Y') = \emptyset$. We also have $A = (A \cap f^{\leftarrow}(X')) \cup (A \cap f^{\leftarrow}(Y'))$.

So A is disconnected, which is a contradiction.

208. (1) Suppose (E, T) is disconnected, say $E = U \cup V$, where U and V are disjoint non-empty T-open subsets of E.

Define the mapping $f : E \to E' = \{0, 1\}$ by setting

$$f(x) = \begin{cases} 0 & \text{if } x \in U \\ 1 & \text{if } x \in V. \end{cases}$$

Since U and V are non-empty, f is a surjection. If T' is the discrete topology on E' we see that f is (T, T')-continuous.

(2) Conversely, suppose there is a (T, T')-continuous surjection f from E onto E'.

Then we have $E = f^{\leftarrow}\{0\} \cup f^{\leftarrow}\{1\}$, $f^{\leftarrow}\{0\} \cap f^{\leftarrow}\{1\} = \emptyset$, $f^{\leftarrow}\{0\} \neq \emptyset$ and $f^{\leftarrow}\{1\} \neq \emptyset$, and $f^{\leftarrow}\{0\}$ and $f^{\leftarrow}\{1\}$ are T-open. So (E, T) is disconnected.

209 (1) If (E, T) is connected and all the sets E_i are non-empty then each projection π_i is a surjection. Since each π_i is (T, T_i)-continuous it follows that each $E_i = \pi_i^{\rightarrow}(E)$ is connected.

(2) Conversely, suppose all the spaces (E_i, T_i) are connected.

If (E, T) is disconnected there exist disjoint non-empty open sets U and V such that $E = U \cup V$.

Let u and v be points of U and V respectively. Since U and V are T-open there are families $(U_i)_{i \in I}$ and $(V_i)_{i \in I}$ such that for each index i in I the sets U_i and V_i are T_i-open, $U_i = V_i = E_i$ for all i not in a certain finite subset J of I and $u \in \prod_{i \in I} U_i \subseteq U$ and $v \in \prod_{i \in I} V_i \subseteq V$. For convenience suppose $J = \{1, 2, \ldots, n\}$.

Since (E, T) is disconnected, E is non-empty; hence all the sets E_i are non-empty. For each index i not in J let p_i be a point in E_i. Now define sets $A_1, \ldots, A_n$ as follows:

$A_1 = \{x \in E : \pi_i(x) = \pi_i(u) \text{ for } 1 < i \leq n \text{ and } \pi_i(x) = p_i \text{ for } i \notin J\},$

$A_n = \{x \in E : \pi_i(x) = \pi_i(v) \text{ for } 1 \leq i < n \text{ and } \pi_i(x) = p_i \text{ for } i \notin J\}$

and for $1 < k < n$

$A_k = \{x \in E : \pi_i(x) = \pi_i(v) \text{ for } 1 \leq i < k, \pi_i(x) = \pi_i(u)$
$\qquad\qquad \text{ for } k < i \leq n \text{ and } \pi_i(x) = p_i \text{ for } i \notin J\}.$

For $k = 1, 2, \ldots, n$ the space (A_k, T_{A_k}) is clearly homeomorphic to (E_k, T_k) and hence is connected. For $k = 1, 2, \ldots, n-1$ the set $A_k \cap A_{k+1}$ is non-empty since it contains the point x_k such that $\pi_i(x_k) = \pi_i(v)$ for $i = 1, \ldots, k$, $\pi_i(x_k) = \pi_i(u)$ for $i = k+1, \ldots, n$ and $\pi_i(x_k) = p_i$ for $i \notin J$. So $A = \bigcup_{1 \leq k \leq n} A_k$ is connected.

Now $A = (A \cap U) \cup (A \cap V)$, $A \cap U \neq \emptyset$, $A \cap V \neq \emptyset$ and hence $A \cap U \cap V \neq \emptyset$. But this is a contradiction since $U \cap V = \emptyset$.

Thus (E, T) is connected.

210. We have

$$C_{E \times E'}(A \times A') = (C_E(A) \times E') \cup (E \times C_{E'}(A'))$$
$$= ((\bigcup_{i \in I} K_i) \times E') \cup (E \times (\bigcup_{j \in J} K'_j)),$$

where $(K_i)_{i \in I}$ is the family of connected components of $C_E(A)$ and $(K'_j)_{j \in J}$ is the family of connected components of $C'_E(A')$.

Since each set $K_i \times E'$ meets each set $E \times K'_j$ and all these sets are connected it follows by a standard argument that $C_{E \times E'}(A \times A')$ is connected.

211. Let (E, T) be a discrete topological space; let x be any point of E.

If A is any subset of E such that $A \supset \{x\}$, A is disconnected. But $\{x\}$ is connected. Hence $\{x\}$ is the connected component of x.

212. Let x be a rational number. Then $\{x\}$ is a connected subset containing x. If A is any subset of $\mathbf{Q}$ such that $A \supset \{x\}$ let y be any point of A distinct from x. There exists an irrational real number z between x and y. Then $A = (A \cap (-\infty, z)) \cup (A \cap (z, \infty))$ is a disconnection of A, so A is not connected. Hence $\{x\}$ is the connected component of x. Thus $\mathbf{Q}$ is totally disconnected.

Let T be the ordinary topology on $\mathbf{R}$. Let U be any $T_{\mathbf{Q}}$-open set containing x. Then $U \supseteq \mathbf{Q} \cap I$ where I is some open interval of $\mathbf{R}$. Hence $U \neq \{x\}$ and hence $T_{\mathbf{Q}}$ is not discrete.

213. For each point x of E let $K(x)$ be the connected component of x. Since $K(x)$ is connected, so is $\operatorname{Cl} K(x)$. Hence $\operatorname{Cl} K(x) \subseteq K(x)$. But of course $K(x) \subseteq \operatorname{Cl} K(x)$. So $\operatorname{Cl} K(x) = K(x)$ and hence $K(x)$ is closed.

The relation R is clearly reflexive.

To show it is symmetric suppose $(x, y) \in R$. Then $y \in K(x)$. Then x belongs to a connected set (namely $K(x)$) which contains y. So $x \in K(y)$. Thus $(y, x) \in R$.

To show that R is transitive suppose $(x, y) \in R$ and $(y, z) \in R$. Then $y \in K(x)$ and $z \in K(y)$. We have $y \in K(x) \cap K(y)$ and hence $K(x) \cup K(y)$ is a connected set which contains x. So $K(x) \cup K(y) \subseteq K(x)$ and hence $z \in K(x)$, i.e. $(x, z) \in R$.

Now let C be any element of E/R, K its connected component.

Suppose K contains more than one point of E/R. Then $\eta^{\leftarrow}(K)$ includes more than one connected component of E and hence is disconnected. So there are T-closed subsets A and B of E such that $\eta^{\leftarrow}(K) \subseteq A \cup B$, $A \cap \eta^{\leftarrow}(K) \neq \emptyset$, $B \cap \eta^{\leftarrow}(K) \neq \emptyset$ and $A \cap B \cap \eta^{\leftarrow}(K) = \emptyset$. Since K is T/R-closed (being a component), $A_1 = A \cap \eta^{\leftarrow}(K)$ and $B_1 = B \cap \eta^{\leftarrow}(K)$ are T-closed and we have $\eta^{\leftarrow}(K) = A_1 \cup B_1$, $A_1 \neq \emptyset$, $B_1 \neq \emptyset$, $A_1 \cap B_1 = \emptyset$. Since $A_1 \cap B_1 = \emptyset$ it follows that A_1 and B_1 are unions of complete R-classes. Hence $A_1 = \eta^{\leftarrow}(\eta^{\rightarrow}(A_1))$ and $B_1 = \eta^{\leftarrow}(\eta^{\rightarrow}(B_1))$. So $\eta^{\rightarrow}(A_1)$ and $\eta^{\rightarrow}(B_1)$ are T/R-closed. Since $K = \eta^{\rightarrow}(A_1) \cup \eta^{\rightarrow}(B_1)$ and $\eta^{\rightarrow}(A_1) \cap \eta^{\rightarrow}(B_1) = \emptyset$ it follows that K is disconnected.

This is a contradiction. So $(E/R, T/R)$ is totally disconnected.

214. Let E be a discrete space with more than one point.

If p is any point of E then $E = \{p\} \cup C_E\{p\}$ shows that (E, T) is disconnected.

If x is any point of E, V any neighbourhood of x, then $\{x\}$ is a connected neighbourhood of x included in V. So (E, T) is locally connected.

215. Let $A = \{(x, y) \in \mathbf{R}^2 : x > 0 \text{ and } y = \sin(1/x)\}$ and let $B = \{(0, 0)\}$.

Then $A = f^{\rightarrow}(\mathbf{R}^+)$ where f is the function given by setting $f(x) = (x, \sin(1/x))$ for all nonzero real numbers x; since $\mathbf{R}^+$ is connected and f is continuous it follows that A is connected. B is connected, and since $(0, 0)$ is an adherent point of A we have $B \cap \text{Cl}\, A \neq \emptyset$. Hence $S = A \cup B$ is connected (Exercise 202).

$(0, 0)$ has no connected neighbourhood included in the $\frac{1}{2}$-ball with centre $(0, 0)$; so S is not locally connected.

216. (1) Suppose (E, T) is locally connected.

Let U be a T-open subset of E. Let K be the connected component (in U) of some point of U. Let t be any point of K.

Since U is open, U is a T-neighbourhood of t. Hence there is a connected T-neighbourhood N of t included in U. Then $N \subseteq K$; so K is a T-neighbourhood of t. Thus K is open.

(2) Conversely, suppose every component of every T-open subset of E is T-open.

Let x be a point of E, V a T-neighbourhood of x. Then V includes a T-open set containing x, U say. The component K of x in U is T-open by hypothesis.

So K is a connected T-neighbourhood of x included in V.

217. Let (E, T) be a locally connected space, R an equivalence relation on E.

Let $\bar{U}$ be a T/R-open subset of E/R, $\bar{x}$ a point of $\bar{U}$, $\bar{K}$ the connected component of $\bar{x}$ in $\bar{U}$. We shall show that $\bar{K}$ is (T/R)-open, i.e. that $\eta^{\leftarrow}(\bar{K})$ is T-open.

Let x be any point of $\eta^{\leftarrow}(\bar{K})$. Then $x \in U = \eta^{\leftarrow}(\bar{U})$, which is T-open. So, since (E, T) is locally connected, the connected component C_x of x in U is T-open. Since C_x is connected and η is $(T, T/R)$-continuous, $\eta^{\rightarrow}(C_x)$ is connected and contains $\eta(x)$, which belongs to $\bar{K}$. So $\eta^{\rightarrow}(C_x) \subseteq \bar{K}$ and consequently $\eta^{\leftarrow}(\bar{K}) \supseteq \eta^{\leftarrow}(\eta^{\rightarrow}(C_x)) \supseteq C_x$. Thus $\eta^{\leftarrow}(\bar{K}) = \bigcup_{x \in \eta^{\leftarrow}(\bar{K})} C_x$, which is T-open.

Hence $\bar{K}$ is (T/R)-open, as required.

218. Let $J = \{i \in I : (E_i, T_i) \text{ is not connected}\}$; by hypothesis J is finite.

Let x be any point of E, V any T-neighbourhood of x. There is a family $(U_i)_{i \in I}$ such that $x \in \prod_{i \in I} U_i \subseteq V$, each U_i is T_i-open, and $U_i = E_i$ for all indices i not in a certain finite subset K. For each index i not in $J \cup K$ set $V_i = E_i$. For each index i in $J \cup K$ the space (E_i, T_i) is locally connected; so there is a connected T_i-neighbourhood V_i of $\pi_i(x)$ included in U_i. Then $\prod_{i \in I} V_i$ is a connected T-neighbourhood of x included in V.

FURTHER READING

The number of books on General Topology is so large that, although it is finite, one is tempted to assert nevertheless that it is uncountable. Over the years, as I have prepared the various courses which have now turned into this book, I have dipped into most of them and have found something of interest in each. But I have studied most carefully N. Bourbaki, *Elements of mathematics: General topology, Parts 1 and 2* (Hermann and Addison-Wesley, 1966) and S. Willard, *General topology* (Addison-Wesley, 1970). These are both much more extensive than the present book and are perhaps more useful as works of reference than as textbooks.

There is a gold-mine of fascinating examples and counterexamples in L.A. Steen and J.A. Seebach, *Counterexamples in topology* (Holt, Rinehart and Winston, 1970).

Finally, for readers who have been infected with the "workbook" approach of this book there is an encyclopædic work in the same style, A.V. Arkhangel'skii and V.I. Ponomarev, *Fundamentals of general topology: Problems and exercises* (Reidel, 1984).

Index